Glaciology and Global Warming

Glaciology and Global Warming

Edited by Michael Waterson

SYRAWOOD
PUBLISHING HOUSE
New York

Published by Syrawood Publishing House,
750 Third Avenue, 9th Floor,
New York, NY 10017, USA
www.syrawoodpublishinghouse.com

Glaciology and Global Warming
Edited by Michael Waterson

International Standard Book Number: 978-1-68286-616-0 (Hardback)

Cataloging-in-Publication Data

Glaciology and global warming / edited by Michael Waterson.
    p. cm.
Includes bibliographical references and index.
ISBN 978-1-68286-616-0
1. Glaciology. 2. Glaciers. 3. Global warming. 4. Greenhouse effect, Atmospheric.
I. Waterson, Michael.
GB2401.7 .G53 2018
551.31--dc23

TABLE OF CONTENTS

PREFACE

The growing pollution has resulted in global warming and melting glaciers. Global warming has become a threat to humans as well as other inhabitants of planet Earth. Glaciology refers to the study of the nature and existing conditions of glaciers, the exact cause of their melting and other natural phenomena concerning ice. The book aims to shed light on some of the unexplored aspects of global warming and glaciology. It is a valuable compilation of topics, ranging from the basic to the most complex theories and principles in the field of glaciology and global warming. This textbook is meant for students who are looking for an elaborate reference text on the subject.

To facilitate a deeper understanding of the contents of this book a short introduction of every chapter is written below:

Chapter 1- The study of glaciers is known as glaciology. It integrates subjects like geology, physical geography, climatology, meteorology, biology, geomorphology and geophysics. Glaciology can be commonly divided into alpine glaciation and continental glaciation. This is an introductory chapter which will introduce briefly all the significant aspects of glaciology.

Chapter 2- Glacier morphology is the study of the formation of glaciers. Glaciers vary in size from large sheets of ice to small cirque glaciers. Plucking, glacial motion, glacial landform, glacial earthquake and glacier mass balance are one of the significant and important topics related to glacier morphology. The following chapter unfolds its crucial aspects in a critical yet systematic manner.

Chapter 3- Anchor ice is a form of ice that can be observed floating in fast flowing rivers during winter season. The other forms of ice discussed are ice cap, ice sheet, ice bridge, ice calving and glacial lake. This chapter provides a plethora of interdisciplinary topics for better comprehension of the bodies of ice.

Chapter 4- An ice age is the period of extreme low temperatures on Earth. The three main types of evidence for ice age are geological, paleontological and chemical. The major ice age periods are Andean-Saharan, Cryogenian, Huronian, Quarternary glaciation and Karoo Ice Age. The aspects elucidated in this chapter are of vital importance, and provide a better understanding of ice age.

Chapter 5- Global warming is the rise in temperatures caused by the release of carbon dioxide and greenhouse gasses into the atmosphere. The rise in temperature caused because of global change has resulted in rising sea levels, melting glaciers, severe droughts, etc. The topics discussed in the chapter are of great importance to broaden the existing knowledge on global warming.

VIII   Preface

Finally, I would like to thank the entire team involved in the inception of this book for their valuable time and contribution. This book would not have been possible without their efforts. I would also like to thank my friends and family for their constant support.

Editor

An Introduction to Glaciology

The study of glaciers is known as glaciology. It integrates subjects like geology, physical geography, climatology, meteorology, biology, geomorphology and geophysics. Glaciology can be commonly divided into alpine glaciation and continental glaciation. This is an introductory chapter which will introduce briefly all the significant aspects of glaciology.

Glaciology

Lateral moraine on a glacier joining the Gorner Glacier, Zermatt, Swiss Alps. The moraine is the high bank of debris in the top left hand quarter of the picture. For more explanation, click on the picture.

Glaciology (from Latin: *glacies*, "frost, ice", and Ancient Greek *logos*, "subject matter"; literally "study of ice") is the scientific study of glaciers, or more generally ice and natural phenomena that involve ice.

Glaciology is an interdisciplinary earth science that integrates geophysics, geology, physical geography, geomorphology, climatology, meteorology, hydrology, biology, and ecology. The impact of glaciers on people includes the fields of human geography and anthropology. The discoveries of water ice on the Moon, Mars, Europa and Pluto add an extraterrestrial component to the field, as in "astroglaciology".

Overview

Areas of study within glaciology include glacial history and the reconstruction of past glaciation. A glaciologist is a person who studies glaciers. A glacial geologist studies glacial deposits and glacial erosive features on the landscape. Glaciology and glacial geology are key areas of polar research. A glacier is an extended mass of ice formed from snow falling and accumulating over a long period of time; they move very slowly, either descending from high mountains, as in valley glaciers, or moving outward from centers of accumulation, as in continental glaciers.

Types

Glacially-carved Yosemite Valley, as seen from a plane

There are two general categories of glaciation which glaciologists distinguish: *alpine glaciation*, accumulations or "rivers of ice" confined to valleys; and *continental glaciation*, unrestricted accumulations which once covered much of the northern continents.

- Alpine - ice flows down the valleys of mountainous areas and forms a tongue of ice moving towards the plains below. Alpine glaciers tend to make the topography more rugged, by adding and improving the scale of existing features such as large ravines called *cirques* and ridges where the rims of two cirques meet called arêtes.

- Continental - an ice sheet found today, only in high latitudes (Greenland/Antarctica), thousands of square kilometers in area and thousands of meters thick. These tend to smooth out the landscapes.

Zones of Glaciers

- Accumulation, where the formation of ice is faster than its removal.

- Wastage or ablation, where the sum of melting and evaporation (sublimation) is greater than the amount of snow added each year.

Movement

Ablation

wastage of the glacier through sublimation, ice melting and iceberg calving.

Ablation zone

> area of a glacier in which the annual loss of ice through ablation exceeds the annual gain from precipitation.

Arête

> an acute ridge of rock where two cirques abut.

Bergschrund

> crevasse formed near the head of a glacier, where the mass of ice has rotated, sheared and torn itself apart in the manner of a geological fault.

Cirque, corrie or cwm

> bowl shaped depression excavated by the source of a glacier.

Creep

> adjustment to stress at a molecular level.

Flow

> movement (of ice) in a constant direction.

Fracture

> brittle failure (breaking of ice) under the stress raised when movement is too rapid to be accommodated by creep. It happens for example, as the central part of a glacier moves faster than the edges.

Horn

> spire of rock, also known as a pyramidal peak, formed by the headward erosion of three or more cirques around a single mountain. It is an extreme case of an arête.

Plucking/Quarrying

> where the adhesion of the ice to the rock is stronger than the cohesion of the rock, part of the rock leaves with the flowing ice.

Tarn

> a post-glacial lake in a cirque.

Tunnel valley

> the tunnel that is formed by hydraulic erosion of ice and rock below an ice sheet

margin. The tunnel valley is what remains of it in the underlying rock when the ice sheet has melted.

Rate of Movement

Movement of the glacier is very slow. It's velocity varies from a few centimeters per day to a few meters per day. The rate of movement depends upon the numbers of factors which are listed below :

- Temperature of the ice

- Gradient of the slope

- Thickness of the glacier

Glacial Deposits

Stratified

Outwash sand/gravel

> from front of glaciers, found on a plain

Kettles

> block of stagnant ice leaves a depression or pit

Eskers

> steep sided ridges of gravel/sand, possibly caused by streams running under stagnant ice

Kames

> stratified drift builds up low steep hills

Varves

> alternating thin sedimentary beds of a proglacial lake. Summer conditions deposit more and coarser material and those of the winter, less and finer.

Unstratified

Till-unsorted

> (glacial flour to boulders) deposited by receding/advancing glaciers, forming moraines, and drumlins

Moraines

> (Terminal) material deposited at the end; (Ground) material deposited as glacier melts; (lateral) material deposited along the sides.

Drumlins

> smooth elongated hills composed of till.

Ribbed moraines

> large subglacial elongated hills transverse to former ice flow.

Glacier

A glacier is a persistent body of dense ice that is constantly moving under its own weight; it forms where the accumulation of snow exceeds its ablation (melting and sublimation) over many years, often centuries. Glaciers slowly deform and flow due to stresses induced by their weight, creating crevasses, seracs, and other distinguishing features. They also abrade rock and debris from their substrate to create landforms such as cirques and moraines. Glaciers form only on land and are distinct from the much thinner sea ice and lake ice that form on the surface of bodies of water.

The Baltoro Glacier in the Karakoram mountains of Pakistan. At 62 kilometres (39 miles) in length, it is one of the longest alpine glaciers on earth.

Ice calving from the terminus of the Perito Moreno Glacier in western Patagonia, Argentina.

The Aletsch Glacier, the largest glacier of the Alps, in Switzerland.

The Quelccaya Ice Cap is the largest glaciated area in the tropics, in Peru.

On Earth, 99% of glacial ice is contained within vast ice sheets in the polar regions, but glaciers may be found in mountain ranges on every continent except Australia, and on a few high-latitude oceanic islands. Between 35°N and 35°S, glaciers occur only in the Himalayas, Andes, Rocky Mountains, a few high mountains in East Africa, Mexico, New Guinea and on Zard Kuh in Iran. Glaciers cover about 10 percent of Earth's land surface. Continental glaciers cover nearly 13,000,000 km² (5×10⁶ sq mi) or about 98 percent of Antarctica›s 13,200,000 km² (5.1×10⁶ sq mi), with an average thickness of 2,100 m (7,000 ft). Greenland and Patagonia also have huge expanses of continental glaciers.

Glacial ice is the largest reservoir of fresh water on Earth. Many glaciers from temperate, alpine and seasonal polar climates store water as ice during the colder seasons and release it later in the form of meltwater as warmer summer temperatures cause the glacier to melt, creating a water source that is especially important for plants, animals and human uses when other sources may be scant. Within high altitude and Antarctic environments, the seasonal temperature difference is often not sufficient to release meltwater.

Because glacial mass is affected by long-term climatic changes, e.g., precipitation, mean temperature, and cloud cover, glacial mass changes are considered among the most sensitive indicators of climate change and are a major source of variations in sea level.

A large piece of compressed ice, or a glacier, appears blue as large quantities of water appear blue. This is because water molecules absorb other colors more efficiently than blue. The other reason for the blue color of glaciers is the lack of air bubbles. Air bubbles, which give a white color to ice, are squeezed out by pressure increasing the density of the created ice.

Etymology and Related Terms

The word *Glaceon* is a loanword from French and goes back, via Franco-Provençal, to the Vulgar Latin *glaciārium*, derived from the Late Latin *glacia*, and ultimately Latin *glaciēs*, meaning "ice". The processes and features caused by or related to glaciers are referred to as glacial. The process of glacier establishment, growth and flow is called glaciation. The corresponding area of study is called glaciology. Glaciers are important components of the global cryosphere.

Types

Classification by Size, Shape, and Behavior

Glaciers are categorized by their morphology, thermal characteristics, and behavior. *Cirque glaciers* form on the crests and slopes of mountains. A glacier that fills a valley is called a *valley glacier*, or alternatively an *alpine glacier* or *mountain glacier*. A large body of glacial ice astride a mountain, mountain range, or volcano is termed an *ice cap* or *ice field*. Ice caps have an area less than 50,000 km² (19,000 sq mi) by definition.

Mouth of the Schlatenkees Glacier near Innergschlöß, Austria

Glacial bodies larger than 50,000 km² (19,000 sq mi) are called *ice sheets* or *continental glaciers*. Several kilometers deep, they obscure the underlying topography. Only nunataks protrude from their surfaces. The only extant ice sheets are the two that cover most of Antarctica and Greenland. They contain vast quantities of fresh water, enough that if both melted, global sea levels would rise by over 70 m (230 ft). Portions of an ice sheet or cap that extend into water are called ice shelves; they tend to be thin with limited slopes and reduced velocities. Narrow, fast-moving sections of an ice sheet are called *ice streams*. In Antarctica, many ice streams drain into large ice shelves. Some drain directly into the sea, often with an ice tongue, like Mertz Glacier.

Sightseeing boat in front of a tidewater glacier, Kenai Fjords National Park, Alaska

Tidewater glaciers are glaciers that terminate in the sea, including most glaciers flowing from Greenland, Antarctica, Baffin and Ellesmere Islands in Canada, Southeast Alaska, and the Northern and Southern Patagonian Ice Fields. As the ice reaches the sea, pieces break off, or calve, forming icebergs. Most tidewater glaciers calve above sea level, which often results in a tremendous impact as the iceberg strikes the water. Tidewater glaciers undergo centuries-long cycles of advance and retreat that are much less affected by the climate change than those of other glaciers.

Classification by Thermal State

Thermally, a *temperate glacier* is at melting point throughout the year, from its surface to its base. The ice of a polar glacier is always below the freezing point from the surface to its base, although the surface snowpack may experience seasonal melting. A *sub-polar glacier* includes both temperate and polar ice, depending on depth beneath the surface and position along the length of the glacier. In a similar way, the thermal regime of a glacier is often described by the temperature at its base alone. A *cold-based glacier* is below freezing at the ice-ground interface, and is thus frozen to the underlying substrate. A *warm-based glacier* is above or at freezing at the interface, and is able to slide at this contact. This contrast is thought to a large extent to govern the ability of a glacier to effectively erode its bed, as sliding ice promotes plucking at rock from the surface below. Glaciers which are partly cold-based and partly warm-based are known as *polythermal*.

Formation

Gorner Glacier in Switzerland

Glaciers form where the accumulation of snow and ice exceeds ablation. The area in which a glacier forms is called a cirque (corrie or cwm) – a typically armchair-shaped geological feature (such as a depression between mountains enclosed by arêtes) – which collects and compresses through gravity the snow which falls into it. This snow collects and is compacted by the weight of the snow falling above it forming névé. Further crushing of the individual snowflakes and squeezing the air from the snow turns it into 'glacial ice'. This glacial ice will fill the cirque until it 'overflows' through a geological weakness or vacancy, such as the gap between two mountains. When the mass of snow and ice is sufficiently thick, it begins to move due to a combination of surface slope, gravity and pressure. On steeper slopes, this can occur with as little as 15 m (50 ft) of snow-ice.

In temperate glaciers, snow repeatedly freezes and thaws, changing into granular ice called firn. Under the pressure of the layers of ice and snow above it, this granular ice fuses into denser and denser firn. Over a period of years, layers of firn undergo further compaction and become glacial ice. Glacier ice is slightly less dense than ice formed from frozen water because it contains tiny trapped air bubbles.

A packrafter passes a wall of freshly-exposed blue ice on Spencer Glacier, in Alaska. Glacial ice acts like a filter on light, and the more time light can spend traveling through ice the bluer it becomes.

Glacial ice has a distinctive blue tint because it absorbs some red light due to an overtone of the infrared OH stretching mode of the water molecule. Liquid water is blue for the same reason. The blue of glacier ice is sometimes misattributed to Rayleigh scattering due to bubbles in the ice.

A glacier cave located on the Perito Moreno Glacier in Argentina.

Structure

A glacier originates at a location called its glacier head and terminates at its glacier foot, snout, or terminus.

Glaciers are broken into zones based on surface snowpack and melt conditions. The ablation zone is the region where there is a net loss in glacier mass. The equilibrium line separates the ablation zone and the accumulation zone; it is the altitude where the amount of new snow gained by accumulation is equal to the amount of ice lost through ablation. The upper part of a glacier, where accumulation exceeds ablation, is called the accumulation zone. In general, the accumulation zone accounts for 60–70% of the glacier's surface area, more if the glacier calves icebergs. Ice in the accumulation zone is deep enough to exert a downward force that erodes underlying rock. After a glacier melts, it often leaves behind a bowl- or amphitheater-shaped depression that ranges in size from large basins like the Great Lakes to smaller mountain depressions known as cirques.

The accumulation zone can be subdivided based on its melt conditions.

1. The dry snow zone is a region where no melt occurs, even in the summer, and the snowpack remains dry.

2. The percolation zone is an area with some surface melt, causing meltwater to percolate into the snowpack. This zone is often marked by refrozen ice lenses, glands, and layers. The snowpack also never reaches melting point.

3. Near the equilibrium line on some glaciers, a superimposed ice zone develops. This zone is where meltwater refreezes as a cold layer in the glacier, forming a continuous mass of ice.

4. The wet snow zone is the region where all of the snow deposited since the end of the previous summer has been raised to 0 °C.

The health of a glacier is usually assessed by determining the glacier mass balance or observing terminus behavior. Healthy glaciers have large accumulation zones, more than 60% of their area snowcovered at the end of the melt season, and a terminus with vigorous flow.

Following the Little Ice Age's end around 1850, glaciers around the Earth have retreated substantially. A slight cooling led to the advance of many alpine glaciers between 1950–1985, but since 1985 glacier retreat and mass loss has become larger and increasingly ubiquitous.

Motion

Shear or herring-bone crevasses on Emmons Glacier (Mount Rainier); such crevasses often form near the edge of a glacier where interactions with underlying or marginal rock impede flow. In this case, the impediment appears to be some distance from the near margin of the glacier.

Glaciers move, or flow, downhill due to gravity and the internal deformation of ice. Ice behaves like a brittle solid until its thickness exceeds about 50 m (160 ft). The pressure on ice deeper than 50 m causes plastic flow. At the molecular level, ice consists of stacked layers of molecules with relatively weak bonds between layers. When the stress on the layer above exceeds the inter-layer binding strength, it moves faster than the layer below.

Glaciers also move through basal sliding. In this process, a glacier slides over the terrain on which it sits, lubricated by the presence of liquid water. The water is created from ice that melts under high pressure from frictional heating. Basal sliding is dominant in temperate, or warm-based glaciers.

Perito Moreno glacier

Fracture Zone and Cracks

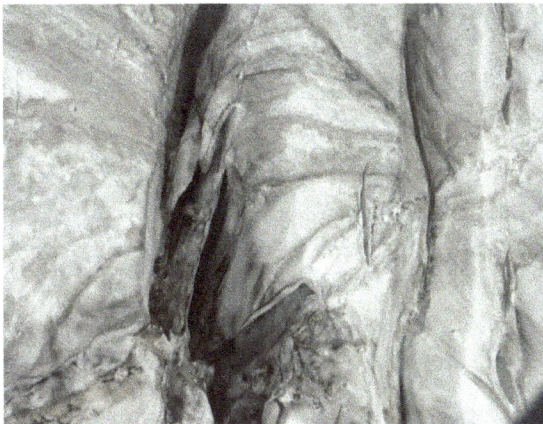
Ice cracks in the Titlis Glacier

The top 50 m (160 ft) of a glacier are rigid because they are under low pressure. This upper section is known as the *fracture zone* and moves mostly as a single unit over the plastically flowing lower section. When a glacier moves through irregular terrain, cracks called crevasses develop in the fracture zone. Crevasses form due to differences in glacier velocity. If two rigid sections of a glacier move at different speeds and directions, shear forces cause them to break apart, opening a crevasse. Crevasses are seldom more than 46 m (150 ft) deep but in some cases can be 300 m (1,000 ft) or even deeper. Beneath this point, the plasticity of the ice is too great for cracks to form. Intersecting crevasses can create isolated peaks in the ice, called seracs.

Crevasses can form in several different ways. Transverse crevasses are transverse to flow and form where steeper slopes cause a glacier to accelerate. Longitudinal crevasses form semi-parallel to flow where a glacier expands laterally. Marginal crevasses form from the edge of the glacier, due to the reduction in speed caused by friction of the valley walls. Marginal crevasses are usually largely transverse to flow. Moving glacier ice can sometimes separate from stagnant ice above, forming a bergschrund. Bergschrunds resemble crevasses but are singular features at a glacier's margins.

Crevasses make travel over glaciers hazardous, especially when they are hidden by fragile snow bridges.

Crossing a crevasse on the Easton Glacier, Mount Baker, in the North Cascades, United States.

Below the equilibrium line, glacial meltwater is concentrated in stream channels. Meltwater can pool in proglacial lakes on top of a glacier or descend into the depths of a glacier via moulins. Streams within or beneath a glacier flow in englacial or sub-glacial tunnels. These tunnels sometimes reemerge at the glacier's surface.

Speed

The speed of glacial displacement is partly determined by friction. Friction makes the ice at the bottom of the glacier move more slowly than ice at the top. In alpine glaciers, friction is also generated at the valley's side walls, which slows the edges relative to the center.

Mean speeds vary greatly, but is typically around 1 m (3 ft) per day. There may be no motion in stagnant areas; for example, in parts of Alaska, trees can establish themselves on surface sediment deposits. In other cases, glaciers can move as fast as 20–30 m (70–100 ft) per day, such as in Greenland's Jakobshavn Isbræ (Greenlandic: *Sermeq Kujalleq*). Velocity increases with increasing slope, increasing thickness, increasing snowfall, increasing longitudinal confinement, increasing basal temperature, increasing meltwater production and reduced bed hardness.

A few glaciers have periods of very rapid advancement called surges. These glaciers exhibit normal movement until suddenly they accelerate, then return to their previous state. During these surges, the glacier may reach velocities far greater than normal speed. These surges may be caused by failure of the underlying bedrock, the pooling of meltwater at the base of the glacier — perhaps delivered from a supraglacial lake — or the simple accumulation of mass beyond a critical "tipping point". Temporary rates up to 90 m (300 ft) per day have occurred when increased temperature or overlying pressure caused bottom ice to melt and water to accumulate beneath a glacier.

In glaciated areas where the glacier moves faster than one km per year, glacial earthquakes occur. These are large scale earthquakes that have seismic magnitudes as high as 6.1. The number of glacial earthquakes in Greenland peaks every year in July, August and September and is increasing over time. In a study using data from January 1993 through October 2005, more events were detected every year since 2002, and twice as many events were recorded in 2005 as there were in any other year. This increase in the numbers of glacial earthquakes in Greenland may be a response to global warming.

Ogives

Ogives are alternating wave crests and valleys that appear as dark and light bands of ice on glacier surfaces. They are linked to seasonal motion of glaciers; the width of one dark and one light band generally equals the annual movement of the glacier. Ogives are formed when ice from an icefall is severely broken up, increasing ablation surface area during summer. This creates a swale and space for snow accumulation in the winter, which in turn creates a ridge. Sometimes ogives consist only of undulations or color bands and are described as wave ogives or band ogives.

Geography

Glaciers are present on every continent and approximately fifty countries, excluding those (Australia, South Africa) that have glaciers only on distant subantarctic island territories. Extensive glaciers are found in Antarctica, Chile, Canada, Alaska, Greenland and Iceland. Mountain glaciers are widespread, especially in the Andes, the Himalayas, the Rocky Mountains, the Caucasus, Scandinavian mountains and the Alps. Mainland Australia currently contains no glaciers, although a small glacier on Mount Kosciuszko was present in the last glacial period. In New Guinea, small, rapidly diminishing, glaciers are located on its highest summit massif of Puncak Jaya. Africa has glaciers on Mount Kilimanjaro in Tanzania, on Mount Kenya and in the Rwenzori Mountains. Oceanic islands with glaciers include Iceland, several of the islands off the coast of Norway including Svalbard and Jan Mayen to the far North, New Zealand and the subantarctic islands of Marion, Heard, Grande Terre (Kerguelen) and Bouvet. During glacial periods of the Quaternary, Taiwan, Hawaii on Mauna Kea and Tenerife also had large alpine glaciers, while the Faroe and Crozet Islands were completely glaciated.

Black ice glacier near Aconcagua, Argentina

The permanent snow cover necessary for glacier formation is affected by factors such as the degree of slope on the land, amount of snowfall and the winds. Glaciers can be found in all latitudes except from 20° to 27° north and south of the equator where the presence of the descending limb of the Hadley circulation lowers precipitation so much that with high insolation snow lines reach above 6,500 m (21,330 ft). Between 19°N and 19°S, however, precipitation is higher and the mountains above 5,000 m (16,400 ft) usually have permanent snow.

Even at high latitudes, glacier formation is not inevitable. Areas of the Arctic, such as Banks Island, and the McMurdo Dry Valleys in Antarctica are considered polar deserts where glaciers cannot form because they receive little snowfall despite the bitter cold. Cold air, unlike warm air, is unable to transport much water vapor. Even during glacial periods of the Quaternary, Manchuria, lowland Siberia, and central and northern Alaska, though extraordinarily cold, had such light snowfall that glaciers could not form.

In addition to the dry, unglaciated polar regions, some mountains and volcanoes in Bolivia, Chile and Argentina are high (4,500 to 6,900 m or 14,800 to 22,600 ft) and cold, but the relative lack of precipitation prevents snow from accumulating into glaciers. This is because these peaks are located near or in the hyperarid Atacama Desert.

Glacial Geology

Glaciers erode terrain through two principal processes: abrasion and plucking.

As glaciers flow over bedrock, they soften and lift blocks of rock into the ice. This process, called plucking, is caused by subglacial water that penetrates fractures in the bedrock and subsequently freezes and expands. This expansion causes the ice to act as a lever that loosens the rock by lifting it. Thus, sediments of all sizes become part of the glacier's load. If a retreating glacier gains enough debris, it may become a rock glacier, like the Timpanogos Glacier in Utah.

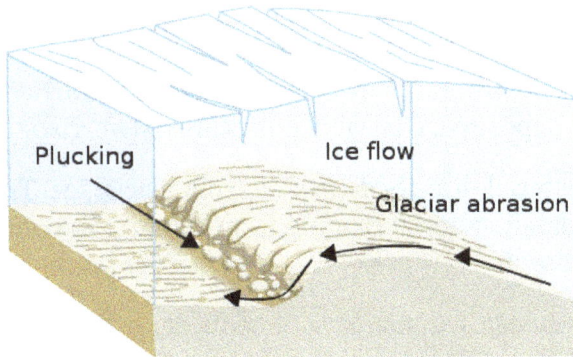

Diagram of glacial plucking and abrasion

Abrasion occurs when the ice and its load of rock fragments slide over bedrock and function as sandpaper, smoothing and polishing the bedrock below. The pulverized rock this process produces is called rock flour and is made up of rock grains between 0.002 and 0.00625 mm in size. Abrasion leads to steeper valley walls and mountain slopes in alpine settings, which can cause avalanches and rock slides, which add even more material to the glacier.

Glacial abrasion is commonly characterized by glacial striations. Glaciers produce these when they contain large boulders that carve long scratches in the bedrock. By mapping the direction of the striations, researchers can determine the direction of the glacier's movement. Similar to striations are chatter marks, lines of crescent-shape depressions in the rock underlying a glacier. They are formed by abrasion when boulders in the glacier are repeatedly caught and released as they are dragged along the bedrock.

The rate of glacier erosion varies. Six factors control erosion rate:

- Velocity of glacial movement
- Thickness of the ice
- Shape, abundance and hardness of rock fragments contained in the ice at the bottom of the glacier
- Relative ease of erosion of the surface under the glacier
- Thermal conditions at the glacier base
- Permeability and water pressure at the glacier base

When the bedrock has frequent fractures on the surface, glacial erosion rates tend to increase as plucking is the main erosive force on the surface; when the bedrock has wide gaps between sporadic fractures, however, abrasion tends to be the dominant erosive form and glacial erosion rates become slow.

Glaciers in lower latitudes tend to be much more erosive than glaciers in higher latitudes, because they have more meltwater reaching the glacial base and facilitate sediment production and transport under the same moving speed and amount of ice.

Material that becomes incorporated in a glacier is typically carried as far as the zone of ablation before being deposited. Glacial deposits are of two distinct types:

- *Glacial till*: material directly deposited from glacial ice. Till includes a mixture of undifferentiated material ranging from clay size to boulders, the usual composition of a moraine.

- *Fluvial and outwash sediments*: sediments deposited by water. These deposits are stratified by size.

Larger pieces of rock that are encrusted in till or deposited on the surface are called "glacial erratics". They range in size from pebbles to boulders, but as they are often moved great distances, they may be drastically different from the material upon which they are found. Patterns of glacial erratics hint at past glacial motions.

Moraines

Glacial moraines above Lake Louise, Alberta, Canada

Glacial moraines are formed by the deposition of material from a glacier and are exposed after the glacier has retreated. They usually appear as linear mounds of till, a non-sorted mixture of rock, gravel and boulders within a matrix of a fine powdery material. Terminal or end moraines are formed at the foot or terminal end of a glacier. Lateral moraines are formed on the sides of the glacier. Medial moraines are formed when two different glaciers merge and the lateral moraines of each coalesce to form a moraine in the middle of the combined glacier. Less apparent are ground moraines, also called *glacial drift*, which often blankets the surface underneath the glacier downslope from the equilibrium line.

The term *moraine* is of French origin. It was coined by peasants to describe alluvial embankments and rims found near the margins of glaciers in the French Alps. In modern geology, the term is used more broadly, and is applied to a series of formations, all of which are composed of till. Moraines can also create moraine dammed lakes.

Drumlins

A drumlin field forms after a glacier has modified the landscape.
The teardrop-shaped formations denote the direction of the ice flow.

Drumlins are asymmetrical, canoe shaped hills made mainly of till. Their heights vary from 15 to 50 meters and they can reach a kilometer in length. The steepest side of the hill faces the direction from which the ice advanced (*stoss*), while a longer slope is left in the ice's direction of movement (*lee*).

Drumlins are found in groups called *drumlin fields* or *drumlin camps*. One of these fields is found east of Rochester, New York; it is estimated to contain about 10,000 drumlins.

Although the process that forms drumlins is not fully understood, their shape implies that they are products of the plastic deformation zone of ancient glaciers. It is believed that many drumlins were formed when glaciers advanced over and altered the deposits of earlier glaciers.

Glacial Valleys, Cirques, Arêtes, and Pyramidal Peaks

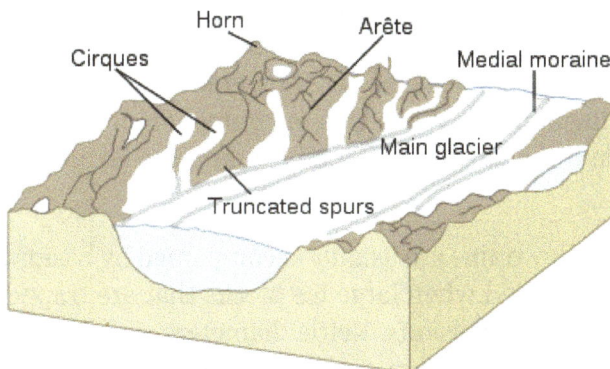

Features of a glacial landscape

Before glaciation, mountain valleys have a characteristic "V" shape, produced by erod-ing water. During glaciation, these valleys are often widened, deepened and smoothed to form a "U"-shaped glacial valley. The erosion that creates glacial valleys truncates any spurs of rock or earth that may have earlier extended across the valley, creating broadly triangular-shaped cliffs called truncated spurs. Within glacial valleys, depres-sions created by plucking and abrasion can be filled by lakes, called paternoster lakes. If a glacial valley runs into a large body of water, it forms a fjord.

Typically glaciers deepen their valleys more than their smaller tributaries. Therefore, when glaciers recede, the valleys of the tributary glaciers remain above the main gla-cier's depression and are called hanging valleys.

At the start of a classic valley glacier is a bowl-shaped cirque, which has escarped walls on three sides but is open on the side that descends into the valley. Cirques are where ice begins to accumulate in a glacier. Two glacial cirques may form back to back and erode their backwalls until only a narrow ridge, called an arête is left. This structure may result in a mountain pass. If multiple cirques encircle a single mountain, they create pointed pyramidal peaks; particularly steep examples are called horns.

Roches Moutonnées

Passage of glacial ice over an area of bedrock may cause the rock to be sculpted into a knoll called a *roche moutonnée,* or "sheepback" rock. Roches moutonnées may be elongated, rounded and asymmetrical in shape. They range in length from less than a meter to several hundred meters long. Roches moutonnées have a gentle slope on their up-glacier sides and a steep to vertical face on their down-glacier sides. The glacier abrades the smooth slope on the upstream side as it flows along, but tears rock frag-ments loose and carries them away from the downstream side via plucking.

Alluvial Stratification

As the water that rises from the ablation zone moves away from the glacier, it carries fine eroded sediments with it. As the speed of the water decreases, so does its capacity to carry objects in suspension. The water thus gradually deposits the sediment as it runs, creating an alluvial plain. When this phenomenon occurs in a valley, it is called a *valley train.* When the deposition is in an estuary, the sediments are known as bay mud.

Outwash plains and valley trains are usually accompanied by basins known as "kettles". These are small lakes formed when large ice blocks that are trapped in alluvium melt and produce water-filled depressions. Kettle diameters range from 5 m to 13 km, with depths of up to 45 meters. Most are circular in shape because the blocks of ice that formed them were rounded as they melted.

Glacial Deposits

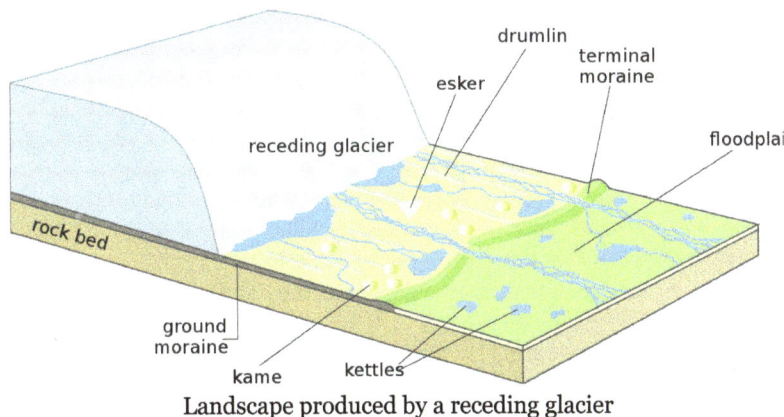

Landscape produced by a receding glacier

When a glacier's size shrinks below a critical point, its flow stops and it becomes stationary. Meanwhile, meltwater within and beneath the ice leaves stratified alluvial deposits. These deposits, in the forms of columns, terraces and clusters, remain after the glacier melts and are known as "glacial deposits".

Glacial deposits that take the shape of hills or mounds are called *kames*. Some kames form when meltwater deposits sediments through openings in the interior of the ice. Others are produced by fans or deltas created by meltwater. When the glacial ice occupies a valley, it can form terraces or kames along the sides of the valley.

Long, sinuous glacial deposits are called *eskers*. Eskers are composed of sand and gravel that was deposited by meltwater streams that flowed through ice tunnels within or beneath a glacier. They remain after the ice melts, with heights exceeding 100 meters and lengths of as long as 100 km.

Loess Deposits

Very fine glacial sediments or rock flour is often picked up by wind blowing over the bare surface and may be deposited great distances from the original fluvial deposition site. These eolian loess deposits may be very deep, even hundreds of meters, as in areas of China and the Midwestern United States of America. Katabatic winds can be important in this process.

Isostatic Rebound

Large masses, such as ice sheets or glaciers, can depress the crust of the Earth into the mantle. The depression usually totals a third of the ice sheet or glacier's thickness. After the ice sheet or glacier melts, the mantle begins to flow back to its original position, pushing the crust back up. This post-glacial rebound, which proceeds very slowly after the melting of the ice sheet or glacier, is currently occurring in measurable amounts in Scandinavia and the Great Lakes region of North America.

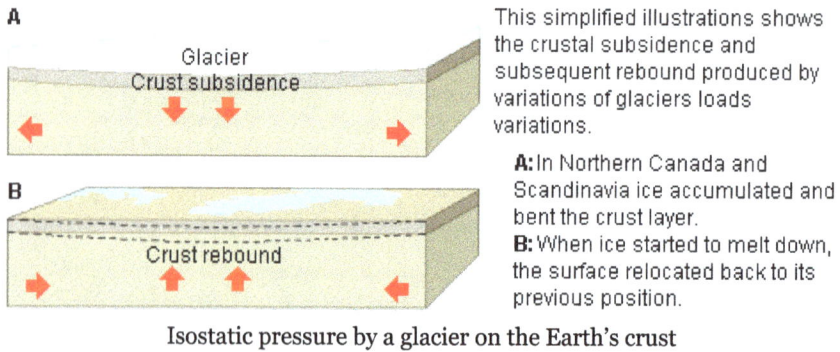

Isostatic pressure by a glacier on the Earth's crust

A geomorphological feature created by the same process on a smaller scale is known as *dilation-faulting*. It occurs where previously compressed rock is allowed to return to its original shape more rapidly than can be maintained without faulting. This leads to an effect similar to what would be seen if the rock were hit by a large hammer. Dilation faulting can be observed in recently de-glaciated parts of Iceland and Cumbria.

On Mars

Northern polar ice cap on Mars

The polar ice caps of Mars show geologic evidence of glacial deposits. The south polar cap is especially comparable to glaciers on Earth. Topographical features and computer models indicate the existence of more glaciers in Mars' past.

At mid-latitudes, between 35° and 65° north or south, Martian glaciers are affected by the thin Martian atmosphere. Because of the low atmospheric pressure, ablation near the surface is solely due to sublimation, not melting. As on Earth, many glaciers are covered with a layer of rocks which insulates the ice. A radar instrument on board the Mars Reconnaissance Orbiter found ice under a thin layer of rocks in formations called lobate debris aprons (LDAs).

The pictures below illustrate how landscape features on Mars closely resemble those on the Earth.

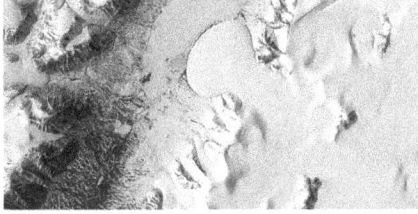

Romer Lake's Elephant Foot Glacier in the Earth's Arctic, as seen by Landsat 8. This picture shows several glaciers that have the same shape as many features on Mars that are believed to also be glaciers.

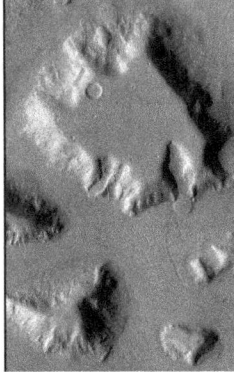

Mesa in Ismenius Lacus quadrangle, as seen by CTX. Mesa has several glaciers eroding it. One of the glaciers is seen in greater detail in the next two images from HiRISE.

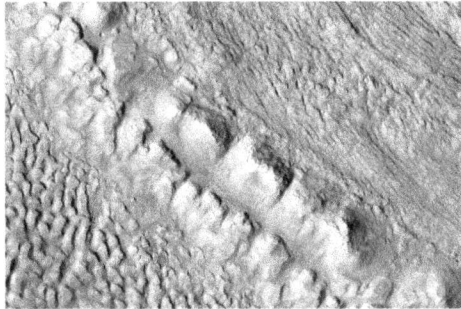

Enlargement of area in rectangle of the previous image. On Earth the ridge would be called the terminal moraine of an alpine glacier. Picture taken with HiRISE under the HiWish program. Image from Ismenius Lacus quadrangle.

Glacier Morphology: A Comprehensive Study

Glacier morphology is the study of the formation of glaciers. Glaciers vary in size from large sheets of ice to small cirque glaciers. Plucking, glacial motion, glacial landform, glacial earthquake and glacier mass balance are one of the significant and important topics related to glacier morphology. The following chapter unfolds its crucial aspects in a critical yet systematic manner.

Glacier Morphology

Franz Josef Glacier in New Zealand

Glacier morphology, or the form a glacier takes, is influenced by temperature, precipitation, topography, and other factors. Types of glaciers range from massive ice sheets, such as the Greenland ice sheet or those in Antarctica, to small cirque glaciers perched on a mountain. Glaciers types can be grouped into two main categories, based on whether ice flow is constrained by the underlying bedrock topography.

Unconstrained

Ice Sheets and Ice Caps

Ice sheets and ice caps cover vast areas and are unconstrained by the underlying topography having a radial flow. The main distinction between the two is the size of their surface, with ice caps covering areas less than 50,000 square kilometers, while ice sheets span larger areas. Ice sheets and ice caps can be classified further.

Vatnajökull ice cap in Iceland

Ice Domes

An ice dome is an upstanding ice surface located in the accumulation zone of the higher altitude portions of an ice cap or ice sheet. Ice domes are nearly symmetrical with a convex or parabolic surface shape. They tend to develop evenly over a land mass that may be either a topographic height or a depression —often reflecting the subglacial topography. In ice sheets, domes may reach a thickness that may exceed 3,000 m, but in ice caps the thickness of the dome is roughly up to several hundred metres. In glaciated islands ice domes are usually the highest point of the ice cap.

An example of an ice dome is Kupol Vostok Pervyy in Alger Island, Franz Josef Land, Russia.

Ice Streams

Ice streams rapidly channel ice flow out to the sea or ocean, where it may feed into an ice shelf. At the margin between ice and water, ice calving takes place, with icebergs breaking off. Ice streams are bounded on the sides by areas of slowly moving ice.

Constrained

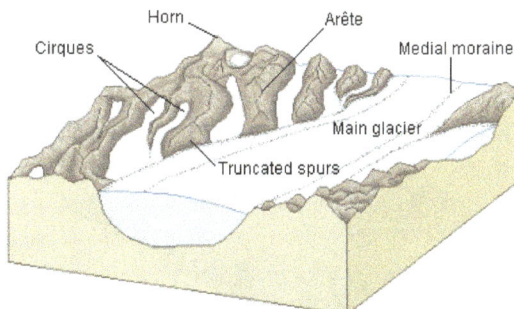

Features of a glacial landscape

Icefield

An icefield covers a relatively large area, usually located in mountainous terrain. The underlying topography controls or influences the form that an icefield takes. Often, nunataks poke through the surface of icefields. Examples of icefields include the Columbia Icefield in the Canadian Rockies and the Northern and Southern Patagonian Ice Field in Chile and Argentina.

Outlet Glaciers

Outlet glaciers are channels of ice that flow out of ice sheets, ice caps or icefields, but are constrained on the sides with exposed bedrock.

Valley Glaciers

Grosser Aletschgletscher, Bernese Alps, Switzerland

Valley glaciers can be outlet glaciers that provide drainage for icefields, icecaps or icesheets and they are also constrained by underlying topography. But they may also form up in mountain ranges as gathering snow turns to ice. Ice-free exposed bedrock and slopes often surround valley glaciers, providing snow and ice from above to accumulate on the glacier via avalanches. True fjords are formed when valley glaciers retreat and sea water fills the void.

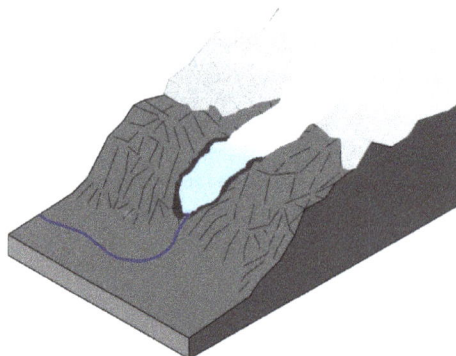

Piedmont Glaciers

Elephant Foot Glacier, a well-known piedmont glacier in Romer Lake,
northeastern Greenland.

Piedmont glaciers are valley glaciers which have spilled out onto relatively flat plains,
where they spread out into bulb-like lobes. The Malaspina Glacier in Alaska is the largest example of this.

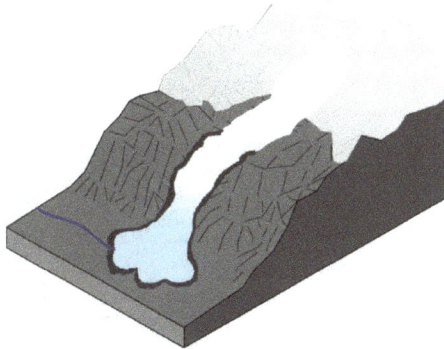

Cirque Glaciers

Lower Curtis Glacier is a cirque glacier in the
North Cascades in the State of Washington.

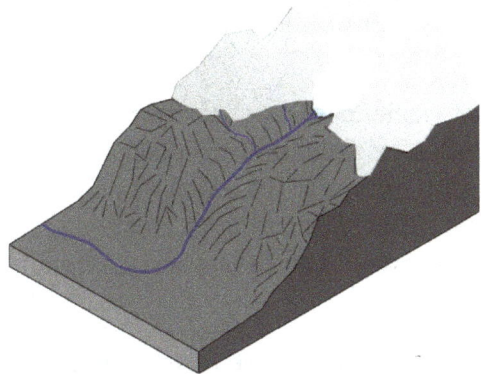

Snow may be situated on the leeward slope of a mountain, where it is sheltered and
accumulates in small depressions. In these depressions, snow persists through summer
months, and is transformed into glacier ice. The glaciers which are built up now, the
cirque glaciers form cirques, bowl-shaped valleys on the side of the mountains.

Glacier Cave

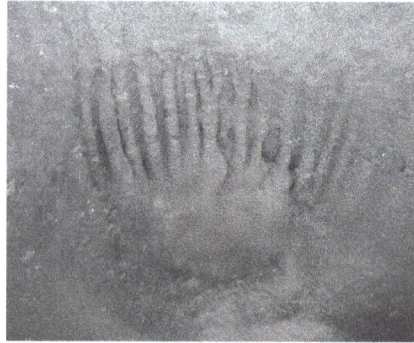

Ice formations in the Titlis glacier cave

A glacier cave is a cave formed within the ice of a glacier. Glacier caves are often called ice caves, but the latter term is properly used to describe bedrock caves that contain year-round ice.

Overview

This glacier cave has been excavated by a hot spring underneath a snow field in south central Iceland, a country where such formations are common due to the high geothermal and volcanic activity, plus the high latitude, cold weather, and frequent snowfall.

Most glacier caves are started by water running through or under the glacier. This water often originates on the glacier's surface through melting, entering the ice at a moulin and exiting at the glacier's snout at base level. Heat transfer from the water can cause sufficient melting to create an air-filled cavity, sometimes aided by solifluction. Air movement can then assist enlargement through melting in summer and sublimation in winter.

Some glacier caves are formed by geothermal heat from volcanic vents or hotsprings beneath the ice. An extreme example is the Kverkfjöll glacier cave in the Vatnajökull glacier in Iceland, measured in the 1980s at 2.8 kilometres (1.7 mi) long with a vertical range of 525 metres (1,722 ft).

Some glacier caves are relatively unstable due to melting and glacial motion, and are subject to localized or complete collapse, as well as elimination by glacial retreat. An example of the dynamic nature of glacier caves is the former Paradise Ice Caves, located on Mt. Rainier in the United States. Known since the early 1900s, the caves were thought to have disappeared altogether in the mid-1940s, yet in 1978 cavers measured 13.25 kilometres (8.23 mi) of passageways in glacier caves there, and it was then considered the longest glacier cave system in the world. The Paradise Ice Caves collapsed and vanished in the 1990s, and the lower lobe of the glacier which once contained the caves has also vanished entirely between 2004 and 2006.

Glacier caves may be used by glaciologists to gain access to the interior of glaciers. The study of glacier caves themselves is sometimes called "glaciospeleology".

Examples

- Mount Rainier (Washington, USA) Two craters on top of a cone on the volcano's summit contain the world's largest volcanic ice-cave system.

- Perito Moreno Glacier (Argentina).

- Titlis (Switzerland).

Plucking (Glaciation)

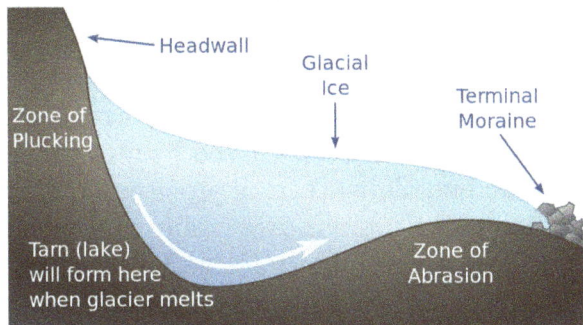

Zone of plucking in the formation of tarns and cirques.

Plucking, also referred to as quarrying, is a glacial phenomenon that is responsible for the erosion and transportation of individual pieces of bedrock, especially large "joint blocks". This occurs in a type of glacier called a "valley glacier". As a glacier moves down a valley, friction causes the basal ice of the glacier to melt and infiltrate joints (cracks) in the bedrock. The freezing and thawing action of the ice enlarges, widens, or causes further cracks in the bedrock as it changes volume across the ice/water phase transition (a form of hydraulic wedging), gradually loosening the rock between the joints. This produces large pieces of rock called joint blocks. Eventually these joint blocks come loose and become trapped in the glacier.

Glacially-plucked granitic bedrock near Mariehamn, Åland Islands.

In this way, plucking has been linked to regelation. Rocks of all sizes can become trapped in the bottom of the glacier. Joint blocks up to three meters have been "plucked" and transported. These entrained rock fragments can also cause abrasion along the subsequent bedrock and walls. Plucking also leads to chatter marks, wedge shaped indentations left on the bedrock or other rock surfaces. Glacial plucking both exploits pre-existing fractures in the bedrock and requires continued fracturing to maintain the cycle of erosion. Glacial plucking is most significant where the rock surface is well jointed or fractured or where it contains exposed bed planes, as this allows meltwater and clasts to penetrate more easily.

Plucking of bedrock also occurs in steep upland rivers, and shares a number of similarities with glacial examples. In such cases, the loosening and detachment of blocks appears to result from a combination of (1) chemical and physical weathering along joints, (2) hydraulic wedging driven by smaller rock fragments getting into existing cracks, (3) crack propagation from stresses caused by impacts of large clasts already in transport by the river, and possibly (4) crack propagation driven by flexing resulting from pressure variation in the overlying water during floods. Loosened blocks are then carried away by fast flowing water during large floods, though the entrainment is believed to be significantly less efficient than the equivalent ability of ice to carry away blocks under glaciers.

Plucking Mechanisms

Glacial plucking is largely dependent on the amount of stress exerted on a clast overlain by glacial ice. This relationship is a balance between the sheer stress exerted on the clast and the normal pressure on the clast by a body of ice. Plucking is increased where there are preexisting fractures in a rock bed. As the glacier slides down a mountain, energy from friction, pressure or geothermal heat causes glacial meltwater to infiltrate the spaces between rocks. This process, known as frost wedging, puts stress on the rock structure as water expands when it freezes. Impacts from large clasts carried in the bedload can cause additional stress to the bedrock. Additionally, plucking can be seen as a positive feedback system in which the increased action of rock removed from the landscape entrained in the glacier causes larger scale fracturing further down the glacier because of a heavier load of force pushing down on the rock bed.

Mechanical Erosion

Glacial plucking is the main mechanism of other small scale mechanical glacial erosion such as striation, abrasion and glacial polishing. The heavier the sediment load, the more extreme the erosion of the downhill landscape. Erosion is largely dependent on the amount of water flow and its velocity, the clast size and hardness with relation to the stability of the slope.

Glacial Striation

A rock that has been subject to glacial erosion will often show a striation pattern in which the rock appears scratched. Long parallel lines will cover the rock and show the appearance of something having been dragged along the top of it. Although striations can form on any sort of rock, they are usually present on more stable bedrock such as quartzite or granite where erosion processes are more readily preserved. Striations, because of their nature of erosion, can also tell geologists the path and movement of the glacier.

Polishing

Glacial polishing is the result of clasts embedded in glacial ice passing over bedrock and grinding down the top of the rock into a smoother surface. The small rocks entrained by plucking act like sandpaper to the downhill slope. This creates an almost mirror like surface in the rock. Polish indicates a more recent process as it is often lost to weathering of the rock surface.

Glacial Till

The joint blocks and rock fragments that are entrained and carried down the slope can be deposited as till. This leads to a whole set of depositional glacial landforms such as moraines, roche moutonnée, glacial erratics and drumlin fields.

Post-glacial Rebound

Post-glacial rebound (also called either isostatic rebound or crustal rebound) is the rise of land masses that were depressed by the huge weight of ice sheets during the last glacial period, through a process known as isostatic depression. Post-glacial rebound and isostatic depression are different parts of a process known as either glacial isostasy, glacial isostatic adjustment, or glacioisostasy. Glacioisostasy is the solid Earth deformation associated with changes in ice mass distribution. The most obvious and direct effects of post-glacial rebound are readily apparent in parts of Northern Eurasia, Northern America, Patagonia, and Antarctica. However, through processes known as *ocean siphoning* and *continental levering*, the effects of post-glacial rebound on sea level are felt globally far from the locations of current and former ice sheets.

This layered beach at Bathurst Inlet, Nunavut is an example of post-glacial rebound after the last Ice Age. Little to no tide helped to form its layer-cake look. Isostatic rebound is still underway here.

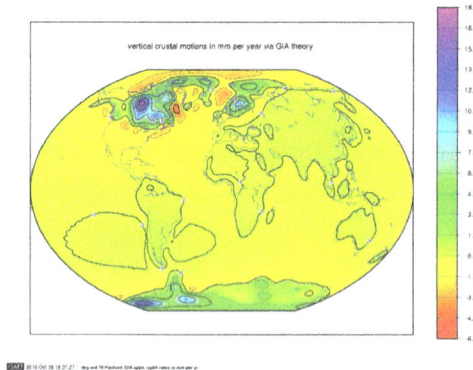

A model of present-day mass change due to post-glacial rebound and the reloading of the ocean basins with seawater. Blue and purple areas indicate rising due to the removal of the ice sheets. Yellow and red areas indicate falling as mantle material moved away from these areas in order to supply the rising areas, and because of the collapse of the forebulges around the ice sheets.

Overview

During the last glacial period, much of northern Europe, Asia, North America, Greenland and Antarctica was covered by ice sheets. The ice was as thick as three kilometres during the last glacial maximum about 20,000 years ago. The enormous weight of this ice caused the surface of the Earth's crust to deform and warp downward, forcing the viscoelastic mantle material to flow away from the loaded region. At the end of each glacial period when the glaciers retreated, the removal of the weight from the depressed land led to slow (and still ongoing) uplift or rebound of the land and the return flow of mantle material back under the deglaciated area. Due to the extreme viscosity of the mantle, it will take many thousands of years for the land to reach an equilibrium level.

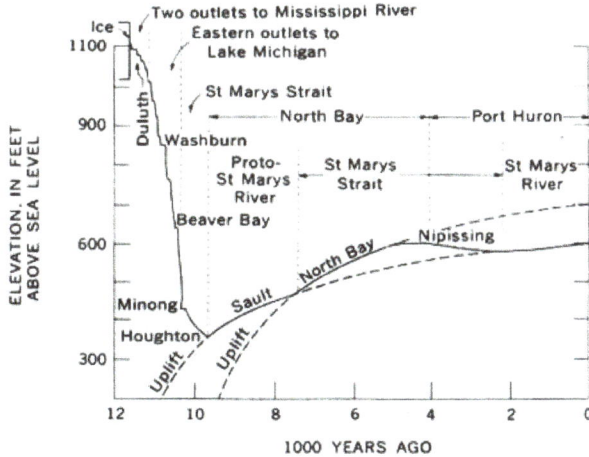

Changes in the elevation of Lake Superior due to glaciation and post-glacial rebound.

The uplift has taken place in two distinct stages. The initial uplift following deglaciation was almost immediate due to the elastic response of the crust as the ice load was removed. After this elastic phase, uplift proceeded by slow viscous flow so the rate of uplift decreased exponentially after that. Today, typical uplift rates are of the order of 1 cm/year or less. In northern Europe, this is clearly shown by the GPS data obtained by the BIFROST GPS network. Studies suggest that rebound will continue for about at least another 10,000 years. The total uplift from the end of deglaciation depends on the local ice load and could be several hundred metres near the centre of rebound.

Recently, the term post-glacial rebound is gradually being replaced by the term glacial isostatic adjustment. This is in recognition that the response of the Earth to glacial loading and unloading is not limited to the upward rebound movement, but also involves downward land movement, horizontal crustal motion, changes in global sea levels and the Earth's gravity field, induced earthquakes, and changes in the rotational motion. An alternate term that is sometimes used is glacial isostasy, because the uplift near the centre of rebound is due to the tendency towards the restoration of isostatic equilibrium (as in the case of isostasy of mountains). Unfortunately, that term gives the wrong impression that isostatic equilibrium is somehow reached, so by appending "adjustment" at the end, the motion of restoration is emphasized.

Effects

Post-glacial rebound produces measurable effects on vertical crustal motion, global sea levels, horizontal crustal motion, gravity field, Earth's rotational motion and state of stress and earthquakes. Studies of glacial rebound give us information about the flow law of mantle rocks and also past ice sheet history. The former is important to the study of mantle convection, plate tectonics and the thermal evolution of the Earth. The latter is important to glaciology, paleoclimate and changes in global sea level. Understanding postglacial rebound is also important to our ability to monitor recent global change.

Vertical Crustal Motion

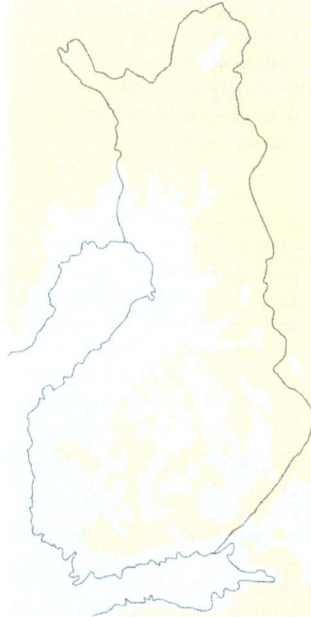

Much of modern Finland is former seabed or archipelago: illustrated are sea levels immediately after the last ice age.

Erratic boulders, U-shaped valleys, drumlins, eskers, kettle lakes, bedrock striations are among the common signatures of the Ice Age. In addition, post-glacial rebound has caused numerous significant changes to coastlines and landscapes over the last several thousand years, and the effects continue to be significant.

In Sweden, Lake Mälaren was formerly an arm of the Baltic Sea, but uplift eventually cut it off and led to its becoming a freshwater lake in about the 12th century, at the time when Stockholm was founded at its outlet. Marine seashells found in Lake Ontario sediments imply a similar event in prehistoric times. Other pronounced effects can be seen on the island of Öland, Sweden, which has little topographic relief due to the presence of the very level Stora Alvaret. The rising land has caused the Iron Age settlement area to recede from the Baltic Sea, making the present day villages on the west coast set back unexpectedly far from the shore. These effects are quite dramatic at the village of Alby, for example, where the Iron Age inhabitants were known to subsist on substantial coastal fishing.

As a result of post-glacial rebound, the Gulf of Bothnia is predicted to eventually close up at Kvarken. The Kvarken is a UNESCO World Natural Heritage Site, selected as a "type area" illustrating the effects of post-glacial rebound and the holocene glacial retreat.

In several other Nordic ports, like Tornio and Pori (formerly at Ulvila), the harbour has had to be relocated several times. Place names in the coastal regions also illustrate the rising land: there are inland places named 'island', 'skerry', 'rock', 'point' and 'sound'. For example, Oulunsalo "island of Oulujoki" is a peninsula, with inland names such as *Koivukari* "Birch Rock", *Santaniemi* "Sandy Cape", and *Salmioja* "the brook of the Sound".

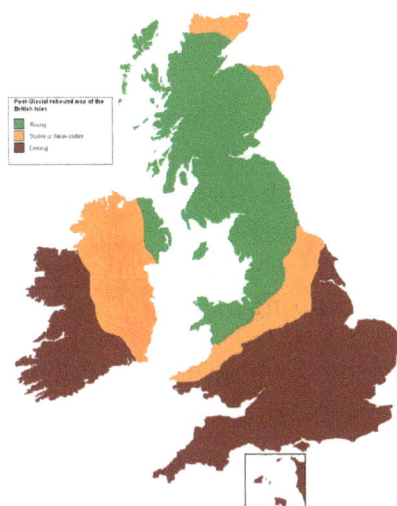

Map of Post Glacial Rebound effects upon the land-level on Ireland and Britain.

In Great Britain, glaciation affected Scotland but not southern England, and the post-glacial rebound of northern Great Britain (up to 10 cm per century) is causing a corresponding downward movement of the southern half of the island (up to 5 cm per century). This will eventually lead to an increased risk of floods in southern England and south-western Ireland.

Since the glacial isostatic adjustment process causes the land to move relative to the sea, ancient shorelines are found to lie above present day sea level in areas that were once glaciated. On the other hand, places in the peripheral bulge area which was uplifted during glaciation now begins to subside. Therefore, ancient beaches are found below present day sea level in the bulge area. The "relative sea level data", which consists of height and age measurements of the ancient beaches around the world, tells us that glacial isostatic adjustment proceeded at a higher rate near the end of deglaciation than today.

The present-day uplift motion in northern Europe is also monitored by a GPS network called BIFROST. Results of GPS data show a peak rate of about 11 mm/year in the north part of the Gulf of Bothnia, but this uplift rate decreases away and becomes negative outside the former ice margin.

In the near field outside the former ice margin, the land sinks relative to the sea. This is the case along the east coast of the United States, where ancient beaches are found submerged below present day sea level and Florida is expected to be submerged in the future. GPS data in North America also confirms that land uplift becomes subsidence outside the former ice margin.

Global Sea Levels

To form the ice sheets of the last Ice Age, water from the oceans evaporated, condensed as snow and was deposited as ice in high latitudes. Thus global sea level fell during glaciation.

The ice sheets at the last glacial maximum were so massive that global sea level fell by about 120 metres. Thus continental shelves were exposed and many islands became connected with the continents through dry land. This was the case between the British Isles and Europe (Doggerland), or between Taiwan, the Indonesian islands and Asia (Sundaland). A sub-continent also existed between Siberia and Alaska that allowed the migration of people and animals during the last glacial maximum.

The fall in sea level also affects the circulation of ocean currents and thus has important impact on climate during the Ice Age.

During deglaciation, the melted ice water returns to the oceans, thus sea level in the ocean increases again. However, geological records of sea level changes show that the redistribution of the melted ice water is not the same everywhere in the oceans. In other words, depending upon the location, the rise in sea level at a certain site may be more than that at another site. This is due to the gravitational attraction between the mass of the melted water and the other masses, such as remaining ice sheets, glaciers, water masses and mantle rocks and the changes in centrifugal potential due to Earth's variable rotation.

Horizontal Crustal Motion

Accompanying vertical motion is the horizontal motion of the crust. The BIFROST GPS network shows that the motion diverges from the centre of rebound. However, the largest horizontal velocity is found near the former ice margin.

The situation in North America is less certain; this is due to the sparse distribution of GPS stations in northern Canada, which is rather inaccessible.

Tilt

The combination of horizontal and vertical motion changes the tilt of the surface. That is, locations farther north rise faster, an effect that becomes apparent in lakes. The bottoms of the lakes gradually tilt away from the direction of the former ice maximum, such that lake shores on the side of the maximum (typically north) recede and the opposite (southern) shores sink. This causes the formation of new rapids and rivers. For example, Lake Pielinen, which is large (90 x 30 km) and oriented perpendicularly to the former ice margin, originally drained through an outlet in the middle of the lake near Nunnanlahti to Lake Höytiäinen. The change of tilt caused Pielinen to burst through the Uimaharju esker at the southwestern end of the lake, creating a new river (Pielisjoki) that runs to the sea via Lake Pyhäselkä to Lake Saimaa. The effects are similar to that concerning seashores, but occur above sea level. Tilting of land will also affect the flow of water in lakes and rivers in the future, and thus important for water resource management planning.

Gravity Field

Ice, water and mantle rocks have mass, and as they move around, they exert a gravitational pull on other masses towards them. Thus, the gravity field, which is sensitive to all mass on the surface and within the Earth, is affected by the redistribution of ice/melted water on the surface of the Earth and the flow of mantle rocks within.

Today, more than 6000 years after the last deglaciation terminated, the flow of mantle material back to the glaciated area causes the overall shape of the Earth to become less oblate. This change in the topography of Earth's surface affects the long-wavelength components of the gravity field.

The changing gravity field can be detected by repeated land measurements with absolute gravimeters and recently by the GRACE satellite mission. The change in long-wavelength components of Earth's gravity field also perturbs the orbital motion of satellites and has been detected by LAGEOS satellite motion.

Vertical Datum

The vertical datum is a theoretical reference surface for altitude measurement and plays vital roles in many human activities, including land surveying and construction of buildings and bridges. Since postglacial rebound continuously deforms the crustal surface and the gravitational field, the vertical datum needs to be redefined repeatedly through time.

State of Stress, Intraplate Earthquakes and Volcanism

According to the theory of plate tectonics, plate-plate interaction results in earthquakes near plate boundaries. However, large earthquakes are found in intraplate environment like eastern Canada (up to M7) and northern Europe (up to M5) which are far away from present-day plate boundaries. An important intraplate earthquake was the magnitude 8 New Madrid earthquake that occurred in mid-continental US in the year 1811.

Glacial loads provided more than 30 MPa of vertical stress in northern Canada and more than 20 MPa in northern Europe during glacial maximum. This vertical stress is supported by the mantle and the flexure of the lithosphere. Since the mantle and the lithosphere continuously respond to the changing ice and water loads, the state of stress at any location continuously changes in time. The changes in the orientation of the state of stress is recorded in the postglacial faults in southeastern Canada. When the postglacial faults formed at the end of deglaciation 9000 years ago, the horizontal principal stress orientation was almost perpendicular to the former ice margin, but today the orientation is in the northeast-southwest, along the direction of seafloor spreading at the Mid-Atlantic Ridge. This shows that the stress due to postglacial rebound had played an important role at deglacial time, but has gradually relaxed so that tectonic stress has become more dominant today.

According to the Mohr–Coulomb theory of rock failure, large glacial loads generally suppress earthquakes, but rapid deglaciation promotes earthquakes. According to Wu & Hasagawa, the rebound stress that is available to trigger earthquakes today is of the order of 1 MPa. This stress level is not large enough to rupture intact rocks but is large enough to reactivate pre-existing faults that are close to failure. Thus, both postglacial rebound and past tectonics play important roles in today's intraplate earthquakes in eastern Canada and southeast US. Generally postglacial rebound stress could have triggered the intraplate earthquakes in eastern Canada and may have played some role in triggering earthquakes in the eastern US including the New Madrid earthquakes of 1811. The situation in northern Europe today is complicated by the current tectonic activities nearby and by coastal loading and weakening.

Increasing pressure due to the weight of the ice during glaciation may have suppressed melt generation and volcanic activities below Iceland and Greenland. On the other hand, decreasing pressure due to deglaciation can increase the melt production and volcanic activities by 20-30 times.

Recent Global Warming

Recent global warming has caused mountain glaciers and the ice sheets in Greenland and Antarctica to melt and global sea level to rise. Therefore, monitoring sea level rise and the mass balance of ice sheets and glaciers allows people to understand more about global warming.

Recent rise in sea levels has been monitored by tide gauges and satellite altimetry (e.g. TOPEX/Poseidon). As well as the addition of melted ice water from glaciers and ice sheets, recent sea level changes are affected by the thermal expansion of sea water due to global warming, sea level change due to deglaciation of the last Ice Age (postglacial sea level change), deformation of the land and ocean floor and other factors. Thus, to understand global warming from sea level change, one must be able to separate all these factors, especially postglacial rebound, since it is one of the leading factors.

Mass changes of ice sheets can be monitored by measuring changes in the ice surface height, the deformation of the ground below and the changes in the gravity field over the ice sheet. Thus ICESat, GPS and GRACE satellite mission are useful for such purpose. However, glacial isostatic adjustment of the ice sheets affect ground deformation and the gravity field today. Thus understanding glacial isostatic adjustment is important in monitoring recent global warming.

One of the possible impacts of global warming-triggered rebound may be more volcanic activity in previously ice-capped areas such as Iceland and Greenland. It may also trigger intraplate earthquakes near the ice margins of Greenland and Antarctica.

Applications

The speed and amount of postglacial rebound is determined by two factors: the viscosity or rheology (i.e., the flow) of the mantle, and the ice loading and unloading histories on the surface of Earth.

The viscosity of the mantle is important in understanding mantle convection, plate tectonics, dynamical processes in Earth, the thermal state and thermal evolution of Earth. However viscosity is difficult to observe because creep experiments of mantle rocks take thousands of years to observe and the ambient temperature and pressure conditions are not easy to attain for a long enough time. Thus, the observations of postglacial rebound provide a natural experiment to measure mantle rheology. Modelling of glacial isostatic adjustment addresses the question of how viscosity changes in the radial and lateral directions and whether the flow law is linear, nonlinear, or composite rheology.

Ice thickness histories are useful in the study of paleoclimatology, glaciology and paleo-oceanography. Ice thickness histories are traditionally deduced from the three types of information: First, the sea level data at stable sites far away from the centers of deglaciation give an eastimate of how much water entered the oceans or equivalently how much ice was locked up at glacial maximum. Secondly, the location and dates of terminal moraines tell us the areal extent and retreat of past ice sheets. Physics of glaciers gives us the theoretical profile of ice sheets at equilibrium, it also says that the thickness and horizontal extent of equilibrium ice sheets are closely related to the basal condition of the ice sheets. Thus the volume of ice locked up is proportional to their instantaneous area. Finally, the heights of ancient beaches in the sea level data and observed land uplift rates (e.g. from GPS or VLBI) can be used to constrain local ice thickness. A popular ice model deduced this way is the ICE5G model. Because the response of the Earth to changes in ice height is slow, it cannot record rapid fluctuation or surges of ice sheets, thus the ice sheet profiles deduced this way only gives the "average height" over a thousand years or so.

Glacial isostatic adjustment also plays an important role in understanding recent global warming and climate change.

Discovery

Before the eighteenth century, it was thought, in Sweden, that sea levels were falling.

On the initiative of Anders Celsius a number of marks were made in rock on different locations along the Swedish coast. In 1765 it was possible to conclude that it was not a lowering of sea levels but an uneven rise of land. In 1865 Thomas Jamieson came up with a theory that the rise of land was connected with the ice age that had been first discovered in 1837. The theory was accepted after investigations by Gerard De Geer of old shorelines in Scandinavia published in 1890.

Legal Status

In areas where the rising of land is seen, it is necessary to define the exact limits of property. In Finland, the "new land" is legally the property of the owner of the water area, not any land owners on the shore. Therefore, if the owner of the land wishes to build a pier over the "new land", he needs the permission of the owner of the (former) water area. The landowner of the shore may redeem the new land at market price. Usually the owner of the water area is the partition unit of the landowners of the shores, a collective holding corporation.

Glaciers on Mars

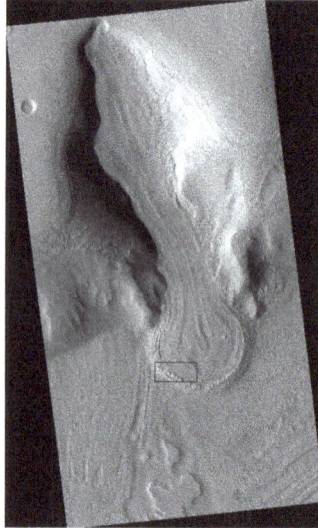

Martian glacier as seen by Hirise. Glacier is moving down valley, then spreading out on plain. Evidence for flow comes from the many lines on surface. The rimming ridges at the end of the glacier are probably moraines. Location is in Protonilus Mensae in Ismenius Lacus quadrangle.

Glaciers, loosely defined as patches of currently or recently flowing ice, are thought to be present across large but restricted areas of the modern Martian surface, and are inferred to have been more widely distributed at times in the past. Lobate convex features on the surface known as viscous flow features and lobate debris aprons, which show the characteristics of non-Newtonian flow, are now almost unanimously regarded as true glaciers. However, a variety of other features on the surface have also been interpreted as directly linked to flowing ice, such as fretted terrain, lineated valley fill, concentric crater fill, and arcuate ridges. A variety of surface textures seen in imagery of the mid-latitudes and polar regions are also thought to be linked to sublimation of glacial ice.

Today, features interpreted as glaciers are largely restricted to latitudes polewards of around 30° latitude. Particular concentrations are found in the Ismenius Lacus quadrangle. Based on our current models of the Martian atmosphere, ice should not however be stable if exposed at the surface in the mid-Martian latitudes. It is thus thought that most glaciers must be covered with a layer of rubble or dust preventing free transfer of water vapor from the subliming ice into the air. This also suggests that in Mars' relatively recent past, its climate must have been different in order to allow the glaciers to grow stably at these latitudes. This provides good independent evidence that the obliquity of Mars has changed significantly in the past, as independently indicated by modelling of Mars' orbital solutions. Evidence for past glaciation also appears on the peaks of several Martian volcanoes in the tropics.

Like glaciers on Earth, glaciers on Mars are not pure water ice. Many are thought to contain substantial proportions of debris, and a substantial number are probably better described as rock glaciers. For many years, largely because of the modeled instability of water ice in the midlatitudes where the putative glacial features were concentrated, it was argued that almost all glaciers were rock glaciers on Mars. However, recent direct observations made by the SHARAD radar instrument on the Mars Reconnaissance Orbiter satellite have confirmed that at least some features are relatively pure ice, and thus, true glaciers. Some authors have also made claims that glaciers of solid carbon dioxide have formed on Mars under certain rare conditions.

Some landscapes look just like glaciers moving out of mountain valleys on Earth. Some have a hollowed out appearance, looking like a glacier after almost all the ice has disappeared. What is left are the moraines—the dirt and debris carried by the glacier. The center is hollowed out because the ice is mostly gone. These supposed alpine glaciers have been called glacier-like forms (GLF) or glacier-like flows (GLF). Glacier-like forms are a later and maybe more accurate term because we cannot be sure the structure is currently moving. Another, more general term sometimes seen in the literature is viscous flow features (VFF).

Radar Studies

Radar studies with the SHAllow RADar (SHARAD) on the Mars Reconnaissance Orbiter showed that lobate debris aprons (LDA) and lineated valley fill (LVF) contain pure water ice covered with a thin layer of rocks that insulated the ice. Ice was found both in the southern hemisphere and in the northern hemisphere. Researchers at the Niels Bohr Institute combined radar observations with ice flow modelling to say that ice in all of the Martian glaciers is equivalent to what could cover the entire surface of Mars with 1.1 meters of ice. The fact that the ice is still there suggests that a thick layer of dust is protecting the ice; the current atmospheric conditions on Mars are such that any exposed water ice would quickly sublimate.

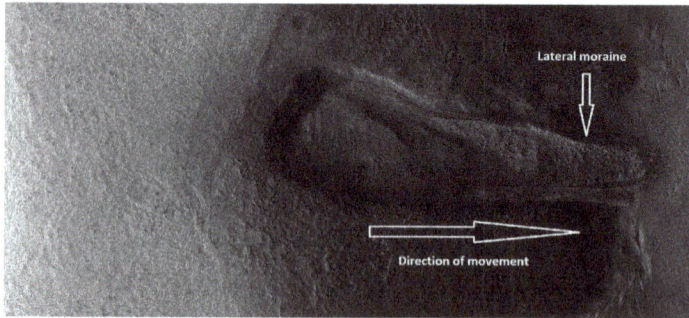

Martian glacier moving down a valley, as seen by HiRISE under HiWish program.

Climate Changes

It is now widely believed that ice accumulated when Mars' orbital tilt was very different from the present (the axis the planet spins on has considerable "wobble," meaning its angle changes over time). A few million years ago, the tilt of the axis of Mars was 45 degrees instead of its present 25 degrees. Its tilt, also called obliquity, varies greatly because its two tiny moons cannot stabilize it like our moon.

Many features on Mars, especially in the Ismenius Lacus quadrangle, are believed to contain large amounts of ice. The most popular model for the origin of the ice is climate change from large changes in the tilt of the planet's rotational axis. At times the tilt has even been greater than 80 degrees Large changes in the tilt explains many ice-rich features on Mars.

Studies have shown that when the tilt of Mars reaches 45 degrees from its current 25 degrees, ice is no longer stable at the poles. Furthermore, at this high tilt, stores of solid carbon dioxide (dry ice) sublimate, thereby increasing the atmospheric pressure. This increased pressure allows more dust to be held in the atmosphere. Moisture in the atmosphere will fall as snow or as ice frozen onto dust grains. Calculations suggest this material will concentrate in the mid-latitudes. General circulation models of the Martian atmosphere predict accumulations of ice-rich dust in the same areas where ice-rich features are found. When the tilt begins to return to lower values, the ice sublimates (turns directly to a gas) and leaves behind a lag of dust. The lag deposit caps the underlying material so with each cycle of high tilt levels, some ice-rich mantle remains behind. Note, that the smooth surface mantle layer probably represents only relative recent material.

Concentric Crater Fill, Lineated Valley Fill, and Lobate Debris Aprons

Several types of landforms have been identified as probably dirt and rock debris covering huge deposits of ice. Concentric crater fill (CCF) contains dozens to hundreds of concentric ridges that are caused by the movements of sometimes hundreds of meter thick accumulations of ice in craters. Lineated valley fill (LVF)are lines of ridges in valleys. These lines may have developed as other glaciers moved down valleys. Some of these glaciers seem to come from material sitting around mesas and buttes. Lobate debris aprons (LDA) is the name given to these glaciers. All of these features that are believed to contain large amounts of ice are found in the mid-latitudes in both the Northern and Southern hemispheres. These areas are sometimes called Fretted terrain because it is sometimes winkled. With the superior resolution of cameras on Mars Global Surveyor (MGS) and MRO, we have found the surface of LDA's, LVF, and CCFs' have a complex tangle of ridges that resemble the surface of the human brain. Wide ridges are called closed-cell brain terrain, and the less common narrow ridges are called open-cell brain terrain. It is thought that the wide closed-cell terrain still contains a core of ice, that when it eventually disappears the center of the wide ridge collapses to produce the narrow ridges of the open-cell brain terrain. Today it is widely accepted that glacier-like forms, lobate debris aprons, lineated valley fill, and concentric fill are all related in that they have the same surface texture. Glacier-like forms in valleys and cirque-like alcoves may coalesce with others to produce lobate debris aprons. When opposing lobate debris aprons converge, linear valley fill results.

Many of these features are found in the Northern hemisphere in parts of a boundary called the Martian dichotomy. The Martian dichotomy is mostly found between 0 and 70 E longitudes. Near this area are regions that are named from ancient names: Deuteronilus Mensae, Protonilus Mensae, and Nilosyrtis Mensae.

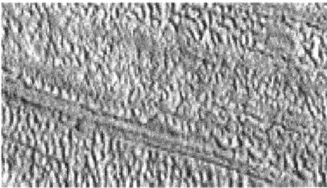

Close, color view of Lineated valley fill in Ismenius Lacus quadrangle, as seen by HiRISE under HiWish program.

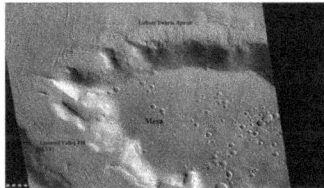

Wide CTX view of mesa showing lobate debris apron (LDA) and lineated valley fill. Both are believed to be debris-covered glaciers. Location is Ismenius Lacus quadrangle.

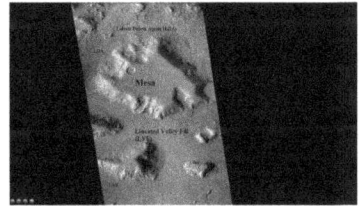

Wide CTX view showing mesa and buttes with lobate debris aprons and lineated valley fill around them. Location is Ismenius Lacus quadrangle.

Open and closed-cell brain terrain, as seen by HiRISE, under HiWish program.

Close-up of lobate debris apron (LDA), as seen by HiRISE under HiWish program.

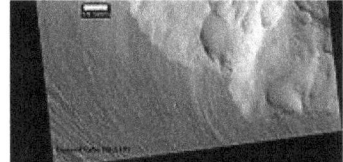

Close-up of lineated valley fill (LVF), as seen by HiRISE under HiWish program.

Tongue-shaped Glaciers

Some of the glaciers flow down mountains and are shaped by obstacles and valleys; they make a sort of tongue shape.

Tongue-shaped glacier, as seen by HiRISE under the HiWish program. Ice may exist in the glacier, even today, beneath an insulating layer of dirt. Location is Hellas quadrangle.

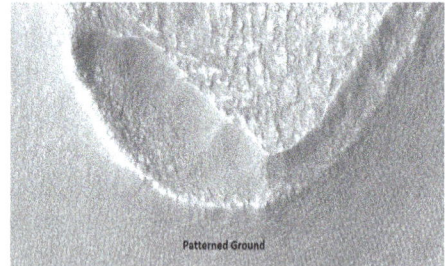

Close-up of tongue-shaped glacier, as seen by HiRISE under the HiWish program. Resolution is about 1 meter, so one can see objects a few meters across in this image.

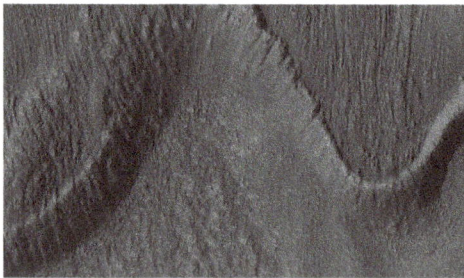

Close-up of the snouts of two glaciers, as seen by HiRISE under the HiWish program.

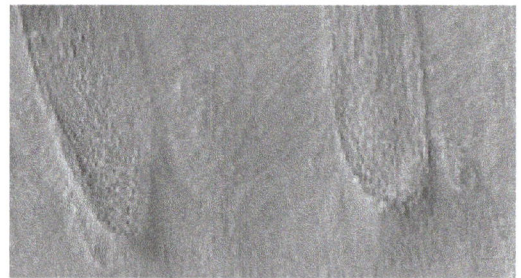

Close view of tongue-shaped flows, as seen by HiRISE under the HiWish program.

Glaciers on Volcanoes

Many supposed glaciers have been observed on some of large Martian volcanoes. Researchers have described glacial deposits on Hecates Tholus, Arsia Mons, Pavonis Mons, and Olympus Mons.

Scientists see evidence that glaciers exist on many of the volcanoes in Tharsis, including Olympus Mons, Ascraeus Mons, and Pavonis Mons. Ceraunius Tholus may have even had its glaciers melt to form some temporary lakes in the past.

Water Source for Future Colonists

Mars has vast glaciers hidden under a layer of rocky debris over wide areas in the mid-latitudes. These glaciers could be large reservoir of life-supporting water on the planet for simple life forms and for future colonists of the Red Planet. Research by John Holt, of the University of Texas at Austin, and others found that one of the features examined is three times larger than the city of Los Angeles and up to 800 m thick, and there are many more.

Some of the glacial-like features were revealed by NASA's Viking orbiters in the 1970s. Since that time glacial-like features have been studied by more and more advanced instruments. Much better data has been received from Mars Global Surveyor, Mars Odyssey, Mars Express, and Mars Reconnaissance Orbiter.

Interactive Mars Map

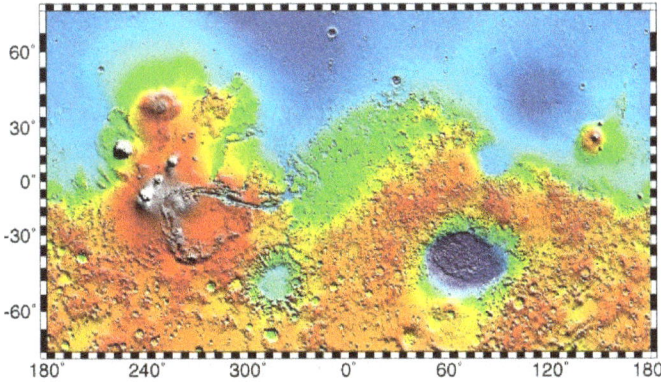

Interactive imagemap of the global topography of Mars
Hover your mouse to see the names of over 25 prominent geographic features, and click to link to them. Coloring of the base map indicates relative elevations, based on data from the Mars Orbiter Laser Altimeter on NASA's Mars Global Surveyor. Reds and pinks are higher elevation (+3 km to +8 km); yellow is 0 km; greens and blues are lower elevation (down to −8 km). Whites (>+12 km) and browns (>+8 km) are the highest elevations. Axes are latitude and longitude; Poles are not shown.

Glacial Motion

This image shows the termini of the glaciers in the Bhutan-Himalaya. Glacial lakes have been rapidly forming on the surface of the debris-covered glaciers in this region during the last few decades. USGS researchers have found a strong correlation between increasing temperatures and glacial retreat in this region.

Glacial motion is the motion of glaciers, which can be likened to rivers of ice. It has played an important role in sculpting many landscapes. Most lakes in the world occupy basins scoured out by glaciers. Glacial motion can be fast (up to 30 m/day, observed on Jakobshavn Isbræ in Greenland) or slow (0.5 m/year on small glaciers or in the center of ice sheets), but is typically around 1 metre/day.

Processes of Motion

Glacier motion occurs from four processes, all driven by gravity: basal sliding, glacial quakes generating fractional movements of large sections of ice, bed deformation, and internal deformation.

- In the case of basal sliding, the entire glacier slides over its bed. This type of motion is enhanced if the bed is soft sediment, if the glacier bed is thawed and if meltwater is prevalent.

- Bed deformation is thus usually limited to areas of sliding. Seasonal melt ponding and penetrating under glaciers shows seasonal acceleration and deceleration of ice flows affecting whole icesheets.

- Some glaciers experience glacial quakes—glaciers "as large as Manhattan and as tall as the Empire State Building, can move 10 meters in less than a minute, a jolt that is sufficient to generate moderate seismic waves." There has been an increasing pattern of these ice quakes - "Quakes ranged from six to 15 per year from 1993 to 2002, then jumped to 20 in 2003, 23 in 2004, and 32 in the first 10 months of 2005." A glacier that is frozen up to its bed does not experience basal sliding.

- Internal deformation occurs when the weight of the ice causes the deformation of ice crystals. This takes place most readily near the glacier bed, where pressures are highest. There are glaciers that primarily move via sliding, glacial quakes, and others that move almost entirely through deformation.

Terminus Movement and Mass Balance

If a glacier's terminus moves forward faster than it melts, the net result is advance. Glacier retreat occurs when more material ablates from the terminus than is replenished by flow into that region.

Glaciologists consider that trends in mass balance for glaciers are more fundamental than the advance or retreat of the termini of individual glaciers. In the years since 1960, there has been a striking decline in the overall volume of glaciers worldwide. This decline is correlated with global warming. As a glacier thins, due to the loss of mass it will slow down and crevassing will decrease.

Landscape and Geology

Studying glacial motion and the landforms that result requires tools from many different disciplines: physical geography, climatology, and geology are among the areas sometime grouped together and called earth science.

During the Pleistocene (the last ice age), huge sheets of ice called continental glaciers advanced over much of the earth. The movement of these continental glaciers created many now-familiar glacial landforms. As the glaciers were expanded, due to their accumulating weight of snow and ice, they crushed and redistributed surface rocks, creating erosional landforms such as striations, cirques, and hanging valleys. Later, when the glaciers retreated leaving behind their freight of crushed rock and sand, depositional landforms were created, such as moraines, eskers, drumlins, and kames. The stone walls found in New England (northeastern United States) contain many glacial erratics, rocks that were dragged by a glacier many miles from their bedrock origin.

At some point, if an Alpine glacier becomes too thin it will stop moving. This will result in the end of any basal erosion. The stream issuing from the glacier will then become clearer as glacial flour diminishes. Lakes and ponds can also be caused by glacial movement. Kettle lakes form when a retreating glacier leaves behind an underground chunk of ice. Moraine-dammed lakes occur when a stream (or snow runoff) is dammed by glacial till.

Glacier Mass Balance

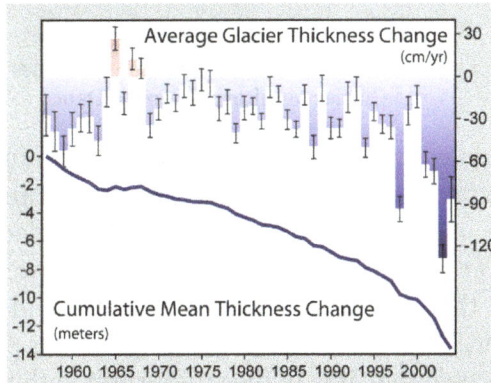

Global glacial mass balance in the last fifty years, reported to the WGMS and NSIDC. The downward trend in the late 1980s is symptomatic of the increased rate and number of retreating glaciers.

Crucial to the survival of a glacier is its mass balance or surface mass balance (SMB), the difference between accumulation and ablation (sublimation and melting). Climate change may cause variations in both temperature and snowfall, causing changes in the surface mass balance. Changes in mass balance control a glacier's long-term behavior and are the most sensitive climate indicators on a glacier. From 1980–2012 the mean cumulative mass loss of glaciers reporting mass balance to the World Glacier Monitoring Service is –16 m. This includes 23 consecutive years of negative mass balances.

Mountain Glacier Changes Since 1970

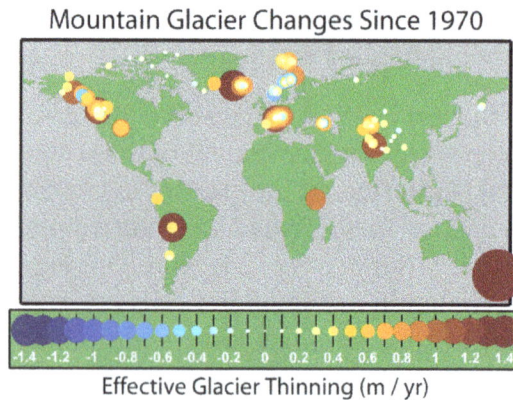

Effective Glacier Thinning (m / yr)

Map of mountain glacier mass balance changes since 1970. Thinning in yellow and red, thickening in blue. The 1970s were a decade of more positive mass balance than the 1980–2004 period.

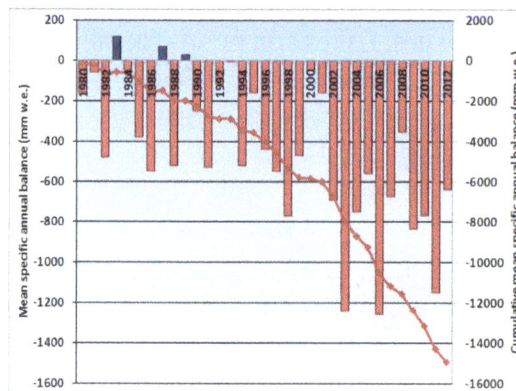

Chart of annual and cumulative glacier mass balance from World Glacier Monitoring Service Data.

A glacier with a sustained negative balance is out of equilibrium and will retreat, while one with a sustained positive balance is out of equilibrium and will advance. Glacier retreat results in the loss of the low elevation region of the glacier. Since higher elevations are cooler than lower ones, the disappearance of the lowest portion of the glacier reduces overall ablation, thereby increasing mass balance and potentially reestablishing equilibrium. However, if the mass balance of a significant portion of the accumulation zone of the glacier is negative, it is in disequilibrium with the local climate. Such a glacier will melt away with a continuation of this local climate. The key symptom of a glacier in disequilibrium is thinning along the entire length of the glacier. For example, Easton Glacier will likely shrink to half its size, but at a slowing rate of reduction, and stabilize at that size, despite the warmer temperature, over a few decades. However, the Grinnell Glacier will shrink at an increasing rate until it disappears. The difference is that the upper section of Easton Glacier remains healthy and snow-covered, while even the upper section of the Grinnell Glacier is bare, melting and has thinned. Small glaciers with shallow slopes such as Grinnell Glacier are most likely to fall into disequilibrium if there is a change in the local climate.

In the case of positive mass balance, the glacier will continue to advance expanding its low elevation area, resulting in more melting. If this still does not create an equilibrium balance the glacier will continue to advance. If a glacier is near a large body of water, especially an ocean, the glacier may advance until iceberg calving losses bring about equilibrium.

Measurement Methods

The Easton Glacier which retreated 255 m from 1990 to 2005 is expected to achieve equilibrium.

Mass Balance

Mass balance is measured by determining the amount of snow accumulated during winter, and later measuring the amount of snow and ice removed by melting in the summer. The difference between these two parameters is the mass balance. If the amount of snow accumulated during the winter is larger than the amount of melted snow and ice during the summer, the mass balance is positive and the glacier has increased in volume. On the other hand, if the melting of snow and ice during the summer is larger than the supply of snow in the winter, the mass balance is negative and the glacier volume decreases. Mass balance is reported in meters of water equivalent. This represents the average thickness gained (positive balance) or lost (negative balance) from the glacier during that particular year.

To determine mass balance in the accumulation zone, snowpack depth is measured using probing, snowpits or crevasse stratigraphy. Crevasse stratigraphy makes use of annual layers revealed on the wall of a crevasse. Akin to tree rings, these layers are due to summer dust deposition and other seasonal effects. The advantage of crevasse stratigraphy is that it provides a two-dimensional measurement of the snowpack layer, not a point measurement. It is also usable in depths where probing or snowpits are not feasible. In temperate glaciers, the insertion resistance of a probe increases abruptly when its tip reaches ice that was formed the previous year. The probe depth is a measure of the net accumulation above that layer. Snowpits dug through the past winters residual snowpack are used to determine the snowpack depth and density. The snowpack's mass balance is the product of density and depth. Regardless of depth measurement technique the observed depth is multiplied by the snowpack density to determine the accumulation in water equivalent. It is necessary to measure the density in the spring as snowpack density varies. Measurement of snowpack density completed at the end of the ablation season yield consistent values for a particular area on temperate alpine glaciers and need not be measured every year. In the ablation zone, ablation measurements are made using stakes inserted vertically into the glacier either at the end of the previous melt season or the beginning of the current one. The length of stake exposed by melting ice is measured at the end of the melt (ablation) season. Most stakes must be replaced each year or even midway through the summer.

Measuring snowpack in a crevasse on the Easton Glacier, North Cascades, USA, the two dimensional nature of the annual layers is apparent.

Measuring snowpack on the Taku Glacier in Alaska, this is a slow and inefficient process, but is very accurate.

Net Balance

Net balance is the mass balance determined between successive mass balance minimums. This is the stratigraphic method focusing on the minima representing a stratigraphic horizon. In the northern mid-latitudes, a glacier's year follows the hydrologic year, starting and ending near the beginning of October. The mass balance minimum is the end of the melt season. The net balance is then the sum of the observed winter balance (bw) normally measured in April or May and summer balance (bs) measured in September or early October.

Measuring snowpack on the Easton Glacier by probing to the previous impenetrable surface, this provides a quick accurate point measurement of snowpack.

Annual Balance

Annual balance is the mass balance measured between specific dates. The mass balance is measured on the fixed date each year, again sometime near the start of October in the mid northern latitudes.

Geodetic Methods

Geodetic methods are an indirect method for the determination of mass balance of glacier. Maps of a glacier made at two different points in time can be compared and the difference in glacier thickness observed used to determine the mass balance over a span of years. This is best accomplished today using Differential Global Positioning System. Sometimes the earliest data for the glacier surface profiles is from images that are used to make topographical maps and digital elevation models. Aerial mapping or photogrammetry is now used to cover larger glaciers and icecaps such found in Antarctica and Greenland, however, because of the problems of establishing accurate ground control points in mountainous terrain, and correlating features in snow and where shading is common, elevation errors are typically not less than 10 m (32 ft). Laser altimetry provides a measurement of the elevation of a glacier along a specific path, e.g., the glacier centerline. The difference of two such measurements is the change in thickness, which provides mass balance over the time interval between the measurements. Again a good method over a span of time but not for annual change detection. The value of geodetic programs is providing an independent check of traditional mass balance work, by comparing the cumulative changes over ten or more years.

Mass Balance Research Worldwide

Mass balance studies have been carried out in various countries worldwide, but have mostly conducted in the Northern Hemisphere due to there being more mid-latitude glaciers in that hemisphere. The World Glacier Monitoring Service annually compiles the mass balance measurements from around the world. From 2002–2006, continuous data is available for only 7 glaciers in the southern hemisphere and 76 glaciers in the Northern Hemisphere. The mean balance of these glaciers was its most negative in any year for 2005/06. The similarity of response of glaciers in western North America indicates the large scale nature of the driving climate change.

Alaska

The Taku Glacier near Juneau, Alaska has been studied by the Juneau Icefield Research Program since 1946, and is the longest continuous mass balance study of any glacier in North America. Taku is the world's thickest known temperate alpine glacier, and experienced positive mass balance between the years 1946 and 1988, resulting in a huge advance. The glacier has since been in a negative mass balance state, which may result in a retreat if the current trends continue. The Juneau Icefield Research Program also has studied the mass balance of the Lemon Creek Glacier since 1953. The glacier has had an average annual balance of –0.44 m per year from 1953–2006, resulting in a mean loss of over 27 m of ice thickness. This loss has been confirmed by laser altimetry.

Austrian Glacier Mass Balance

The mass balance of Hintereisferner and Kesselwandferner glaciers in Austria have been continuously monitored since 1952 and 1965 respectively. Having been continuously measured for 55 years, Hintereisferner has one of the longest periods of continuous study of any glacier in the world, based on measured data and a consistent method of evaluation. Currently this measurement network comprises about 10 snow pits and about 50 ablation stakes distributed across the glacier. In terms of the cumulative specific balances, Hintereisferner experienced a net loss of mass between 1952 and 1964, followed by a period of recovery to 1968. Hintereisferner reached an intermittent minimum in 1976, briefly recovered in 1977 and 1978 and has continuously lost mass in the 30 years since then. Total mass loss has been 26 m since 1952 Sonnblickkees Glacier has been measured since 1957 and the glacier has lost 12 m of mass, an average annual loss of –0.23 m per year.

New Zealand

Glacier mass balance studies have been ongoing in New Zealand since 1957. Tasman Glacier has been studied since then by the New Zealand Geological Survey and later by the Ministry of Works, measuring the ice stratigraphy and overall movement. However, even earlier fluctuation patterns were documented on the Franz Josef and Fox Glaciers in 1950. Other glaciers on the South Island studied include Ivory Glacier since 1968, while on the North Island, glacier retreat and mass balance research has been conducted on the glaciers on Mount Ruapehu since 1955. On Mount Ruapehu, permanent photographic stations allow repeat photography to be used to provide photographic evidence of changes to the glaciers on the mountain over time.

An aerial photographic survey of 50 glaciers in the South Island has been carried out for most years since 1977. The data was used to show that between 1976 and 2005 there was a 10% loss in glacier volume.

North Cascade Glacier Mass Balance Program

The North Cascade Glacier Climate Project measures the annual balance of 10 glaciers, more than any other program in North America, to monitor an entire glaciated mountain range, which was listed as a high priority of the National Academy of Sciences in 1983. These records extend from 1984–2008 and represent the only set of records documenting the mass balance changes of an entire glacier clad range. North Cascade glaciers annual balance has averaged –0.48 m/a from 1984–2008, a cumulative thickness loss of over 13 m or 20–40% of their total volume since 1984 due to negative mass balances. The trend in mass balance is becoming more negative which is fueling more glacier retreat and thinning.

Norway Mass Balance Program

Norway maintains the most extensive mass balance program in the world and is largely funded by the hydropower industry. Mass balance measurements are currently (2012) performed on fifteen glaciers in Norway. In southern Norway six of the glaciers have been measured continuously since 1963 or earlier, and they constitute a west-east profile reaching from the maritime Ålfotbreen Glacier, close to the western coast, to the continental Gråsubreen Glacier, in the eastern part of Jotunheimen. Storbreen Glacier in Jotunheimen has been measured for a longer period of time than any other glacier in Norway, starting in 1949, while Engabreen Glacier at Svartisen has the longest series in northern Norway (starting in 1970). The Norwegian program is where the traditional methods of mass balance measurement were largely derived.

Sweden Storglaciären

The Tarfala research station in the Kebnekaise region of northern Sweden is operated by Stockholm University. It was here that the first mass balance program was initiated immediately after World War II, and continues to the present day. This survey was the initiation of the mass balance record of Storglaciären Glacier, and constitutes the longest continuous study of this type in the world. Storglaciären has had a cumulative negative mass balance from 1946–2006 of –17 m. The program began monitoring the Rabots Glaciär in 1982, Riukojietna in 1985, and Mårmaglaciären in 1988. All three of these glaciers have had a strong negative mass balance since initiation.

Iceland Glacier Mass Balance

Glacier mass balance is measured once or twice annually on numerous stakes on the several ice caps in Iceland by the National Energy Authority. Regular pit and stake mass-balance measurements have been carried out on the northern side of Hofsjökull since 1988 and likewise on the Þrándarjökull since 1991. Profiles of mass balance (pit and stake) have been established on the eastern and south-western side of Hofsjökull since 1989. Similar profiles have been assessed on the Tungnaárjökull, Dyngjujökull, Köldukvíslarjökull and Brúarjökull outlet glaciers of Vatnajökull since 1992 and the Eyjabakkajökull outlet glacier since 1991.

Swiss Mass Balance Program

Temporal changes in the spatial distribution of the mass balance result primarily from changes in accumulation and melt along the surface. As a consequence, variations in the mass of glaciers reflect changes in climate and the energy fluxes at the Earth's surface. The Swiss glaciers Gries in the central Alps and Silvretta in the eastern Alps, have been measured for many years. The distribution of seasonal accumulation and ablation rates are measured in-situ. Traditional field methods are combined with remote sensing techniques to track changes in mass, geometry and the flow behaviour of the two glaciers. These investigations contribute to the Swiss Glacier Monitoring Network and the International network of the World Glacier Monitoring Service (WGMS).

United States Geological Survey (USGS)

The USGS operates a long-term "benchmark" glacier monitoring program which is used to examine climate change, glacier mass balance, glacier motion, and stream runoff. This program has been ongoing since 1965 and has been examining three glaciers in particular. Gulkana Glacier in the Alaska Range and Wolverine Glacier in the Coast Ranges of Alaska have both been monitored since 1965, while the South Cascade Glacier in Washington State has been continuously monitored since the International Geophysical Year of 1957. This program monitors one glacier in each of these mountain ranges, collecting detailed data to understand glacier hydrology and glacier climate interactions.

Geological Survey of Canada-glaciology Section (GSC)

The GSC operates Canada's Glacier-Climate Observing System as part of its Climate Change Geoscience Program. With its University partners, it conducts monitoring and research on glacier-climate changes, water resources and sea level change using a network of reference observing sites located in the Cordillera and the Canadian Arctic Archipelago. This network is augmented with remote sensing assessments of regional glacier changes. Sites in the Cordillera include the Helm, Place, Andrei, Kaskakwulsh, Haig, Peyto, Ram River, Castle Creek, Kwadacha and Bologna Creek Glaciers; in the Arctic Archipelago include the White, Baby and Grise Glaciers and the Devon, Meighen, Melville and Agassiz Ice Caps. GSC reference sites are monitored using the standard stake based glaciological method (stratigraphic) and periodic geodetic assessments using airborne lidar. Detailed information, contact information and database available here: Helm Glacier (–33 m) and Place Glacier (–27 m) have lost more than 20% of their entire volume, since 1980, Peyto Glacier (–20 m) is close to this amount. The Canadian Arctic White Glacier has not been as negative at (–6 m) since 1980.

Bolivia Mass Balance Network

The glacier monitoring network in Bolivia, a branch of the glacio-hydrological system of observation installed throughout the tropical Andes mountains by IRD and partners since 1991, has monitored mass balance on Zongo (6000 m asl), Chacaltaya (5400 m asl) and Charquini glaciers (5380 m asl). A system of stakes has been used, with frequent field observations, as often as monthly. These measurements have been made in concert with energy balance to identify the cause of the rapid retreat and mass balance loss of these tropical glaciers.

PTAA-mass Balance Model

A recently developed glacier balance model based on Monte Carlo principals is a promising supplement to both manual field measurements and geodetic methods of measuring mass balance using satellite images. The PTAA (precipitation-temperature-area-altitude) model requires only daily observations of precipitation and temperature collected at usually low-altitude weather stations, and the area-altitude distribution of the glacier. Output are daily snow accumulation (Bc) and ablation (Ba) for each altitude interval, which is converted to mass balance by $Bn = Bc - Ba$. Snow Accumulation (Bc) is calculated for each area-altitude interval based on observed precipitation at one or more lower altitude weather stations located in the same region as the glacier and three coefficients that convert precipitation to snow accumulation. It is necessary to use established weather stations that have a long unbroken records so that annual means and other statistics can be determined. Ablation (Ba) is determined from temperature observed at weather stations near the glacier. Daily maximum and minimum temperatures are converted to glacier ablation using twelve coefficients.

The fifteen independent coefficients that are used to convert observed temperature and precipitation to ablation and snow accumulation apply a simplex optimizing procedure. The simplex automatically and simultaneously calculates values for each coefficient using Monte Carlo principals that rely on random sampling to obtain numerical results. Similarly, the PTAA model makes repeated calculations of mass balance, minutely re-adjusting the balance for each iteration.

The PTAA model has been tested for eight glaciers in Alaska, Washington, Austria and Nepal. Calculated annual balances are compared with measured balances for approximately 60 years for each of five glaciers. The Wolverine and Gulkana in Alaska, Hintereisferner, Kesselwandferner and Vernagtferner in Austria. It has also been applied to the Langtang Glacier in Nepal. Results for these tests are shown on the GMB (glacier mass balance) website at ptaagmb.com. Linear regressions of model versus manual balance measurements are based on a split-sample approach so that the calculated mass balances are independent of the temperature and precipitation used to calculate the mass balance.

Regression of model versus measured annual balances yield R^2 values of 0.50 to 0.60. Application of the model to Bering Glacier in Alaska demonstrated a close agreement with ice volume loss for the 1972–2003 period measured with the geodetic method. Determining the mass balance and runoff of the partially debris-covered Langtang Glacier in Nepal demonstrates an application of this model to a glacier in the Himalayan Range.

Correlation between ablation of glaciers in the Wrangell Range in Alaska and global temperatures observed at 7000 weather stations in the Northern Hemisphere indicates that glaciers are more sensitive to the global climate than are individual temperature stations, which do not show similar correlations.

Validation of the model to demonstrate the response of glaciers in Northwestern United States to future climate change is shown in a hierarchical modeling approach. Climate downscaling to estimate glacier mass using the PTAA model is applied to determine the balance of the Bering and Hubbard Glaciers and is also validated for the Gulkana, a USGS benchmark glacier.

Glacial Landform

Yosemite Valley from an airplane, showing the U-shape.

Glacially-plucked granitic bedrock near Mariehamn, Åland Islands.

Glacial landforms are landforms created by the action of glaciers. Most of today's glacial landforms were created by the movement of large ice sheets during the Quaternary glaciations. Some areas, like Fennoscandia and the southern Andes, have extensive occurrences of glacial landforms; other areas, such as the Sahara, display very old fossil glacial landforms.

Erosional Landforms

As the glaciers expanded, due to their accumulating weight of snow and ice, they crush and abrade scoured surface rocks and bedrock. The resulting erosional landforms include striations, cirques, glacial horns, arêtes, trim lines, U-shaped valleys, roches moutonnées, overdeepenings and hanging valleys.

- Cirque: Starting location for mountain glaciers

- Cirque stairway: a sequence of cirques

- U-shaped or trough valley: U-shaped valleys are created by mountain glaciers. When filled with ocean water so as to create an inlet, these valleys are called fjords.

- Arête: spiky high land between two glaciers, if the glacial action erodes through, a *spillway* (or col) forms.

- Valley step: an abrupt change in the longitudinal slope of a glacial valley

Depositional Landforms

Later, when the glaciers retreated leaving behind their freight of crushed rock and sand (glacial drift), they created characteristic depositional landforms. Examples include glacial moraines, eskers, and kames. Drumlins and ribbed moraines are also landforms left behind by retreating glaciers. The stone walls of New England contain many glacial erratics, rocks that were dragged by a glacier many miles from their bedrock origin.

- Esker: Built up bed of a subglacial stream.

- Kame: Irregularly shaped mound.

- Moraine: Feature can be terminal (at the end of a glacier), lateral (along the sides of a glacier), or medial (formed by the emerger of lateral moraines from contributary glaciers).

- Outwash fan: Braided stream flowing from the front end of a glacier.

Glacial Lakes and Ponds

Lakes and ponds may also be caused by glacial movement. Kettle lakes form when a retreating glacier leaves behind an underground or surface chunk of ice that later melts to form a depression containing water. Moraine-dammed lakes occur when glacial debris dam a stream (or snow runoff). Jackson Lake and Jenny Lake in Grand Teton National Park are examples of moraine-dammed lakes, though Jackson Lake is enhanced by a man-made dam.

- Kettle lake: Depression, formed by a block of ice separated from the main glacier, in which the lake forms.

- Tarn: A lake formed in a cirque by overdeepening.

- Paternoster lake: A series of lakes in a glacial valley, formed when a stream is dammed by successive recessional moraines left by an advancing or retreating glacier.

- Glacial Lake: A lake that formed between the front of a glacier and the last recessional moraine.

Ice Features

Apart from the landforms left behind by glaciers, glaciers themselves may be striking features of the terrain, particularly in the polar regions of the earth. Notable examples include valley glaciers where glacial flow is restricted by the valley walls, crevasses in the upper section of glacial ice, and icefalls—the ice equivalent of waterfalls.

References

- Holt, J.W.; et al. (2008). "Radar sounding evidence for buried glaciers in the southern mid-latitudes of Mars". Science. 322: 1235–1238. PMID 19023078. doi:10.1126/science.1164246

- Mitrovica, J.X.; W.R. Peltier (1993). "Present-day secular variations in zonal harmonics of the Earth's geopotential". Journal of Geophysical Research. 98: 4509–4526. Bibcode:1993JGR....98.4509M. doi:10.1029/92JB02700

- Nanna Gunnarsdóttir (n.d.). "Caves in Iceland". Guide to Island, a collaboration of more than 300 travel companies and individuals. Retrieved 20 January 2015

- Harbor, Jonathan (2011). Encyclopedia of Snow, Ice and Glaciers (PDF). Springer. pp. 332–340. ISBN 978-90-481-2641-5

- Lambeck, K.; C. Smither; P. Johnston (July 1998). "Sea-level change, glacial rebound and mantle viscosity for northern Europe". Geophysical Journal International. 134 (1): 102–144. Bibcode:1998GeoJI.134..102L. doi:10.1046/j.1365-246x.1998.00541.x

- Wu, P.; W.R.Peltier (1984). "Pleistocene deglaciation and the earth's rotation: a new analysis". Geophysical Journal of the Royal Astronomical Society. 76: 753–792. doi:10.1111/j.1365-246X.1984.tb01920.x

- Gray, Louise (7 October 2009). "England is sinking while Scotland rises above sea levels, according to new study". Telegraph. Retrieved 10 April 2012

- Milne, G.A., and Shennan, I. (2013) Isostasy: glaciation-induced sea-level change. In Encyclopedia of Quaternary Science. volume 3, Elsevier, Oxford, pp. 452-459. ISBN 978-0-444-53643-3

- Jull, M.; D. McKenzie (1996). "The effect of deglaciation on mantle melting beneath Iceland". Journal of Geophysical Research. 101: 21,815–21,828. Bibcode:1996JGR...10121815J. doi:10.1029/96jb01308

- Yoder, C. F.; et al. (1983). "J2-dot from Lageos and the non-tidal acceleration of earth rotation". Nature. 303 (5920): 757–762. Bibcode:1983Natur.303..757Y. doi:10.1038/303757a0

- Van der Wal, W.; et al. (2010). "Sea levels and uplift rate from composite rheology in glacial isostatic adjustment modeling". Journal of Geodynamics. 50: 38–48. Bibcode:2010JGeo...50...38V. doi:10.1016/j.jog.2010.01.006

- Mauri S. Pelto (Nichols College). "The Disequilibrium of North Cascade, Washington Glaciers 1984–2004". In "Hydrologic Processes". Retrieved February 14, 2006

- "The Surface of Mars" Series: Cambridge Planetary Science (No. 6) ISBN 978-0-511-26688-1 Michael H. Carr, United States Geological Survey, Menlo Park

- Lucchitta, Baerbel K (1984). "Ice and debris in the fretted terrain, Mars". Journal of Geophysical Research: Solid Earth. 89 (S02): B409–B418. Bibcode:1984JGR....89..409L. doi:10.1029/jb089is02p0b409

- Karlsson, N.; Schmidt, L.; Hvidberg, C. (2015). "Volume of Martian mid-latitude glaciers from radar observations and ice-flow modelling". Geophysical Research Letters. 42: 2627–2633. doi:10.1002/2015GL063219

- Colaprete, Anthony, and Bruce M. Jakosky. "Ice flow and rock glaciers on Mars." Journal of Geophysical Research: Planets 103.E3 (1998): 5897-5909

An Integrated Study of Bodies of Ice

Anchor ice is a form of ice that can be observed floating in fast flowing rivers during winter season. The other forms of ice discussed are ice cap, ice sheet, ice bridge, ice calving and glacial lake. This chapter provides a plethora of interdisciplinary topics for better comprehension of the bodies of ice.

Anchor Ice

Anchor ice growing on the sea floor in McMurdo Sound, Antarctica.

Anchor ice is defined by the World Meteorological Organization as "submerged ice attached or anchored to the bottom, irrespective of the nature of its formation". It may also be called bottom-fast ice. Anchor ice is most commonly observed in fast-flowing rivers during periods of extreme cold, at the mouths of rivers flowing into very cold seawater, in the shallow sub or intertidal during or after storms when the air temperature is below the freezing point of the water, and the subtidal in the Antarctic along ice shelves or near floating glacier tongues, and in shallow lakes.

Types and Formation

In Rivers

Anchor ice will generally form in fast-flowing rivers during periods of extreme cold. Due to the motion of the water, ice cover may not form consistently, and the water will quickly reach its freezing point due to mixing and contact with the atmosphere. Ice platelets generally form very quickly in the water column and on submerged objects once conditions are optimal for anchor ice formation. Anchor ice in rivers tends to be composed of numerous small crystals adhering to each other in small flocculent masses. Anchor ice in rivers can seriously disrupt hydro-electric power plants by significantly reducing flow or stopping turbines completely.

Anchor ice on the ground of river Saale in Jena, Germany.

Another form of anchor ice may be observed at the mouths of Arctic rivers where fresh water seeps out of the river bed into the ocean up through the sediment. Anchor ice forms if the seawater is below the freezing point of the river water.

In Lakes

Shallow tundra lakes may feature anchor ice with a specific behavior. Lakes in the southwestern part of Nunavut, Canada typically freeze down to the bottom when the water level is low. On some cases spring meltwater flows into the lake under the ice cover, which becomes domed leaving a depressed "racetrack" ring around the shore where meltwater accumulates as well. The ice cover remains bottom-fast until the buoyancy force exceeds the freezing bond. At the latter moment the ice cover abruptly breaks off the bottom to form a flat sheet. In other cases the anchor ice becomes completely submerged into the meltwater and holes may be melted throughout the ice sheet. When the sheet finally lifts off the bottom, the meltwater accumulated at the surface is jetted through these holes with enough force to create small craters in the lake bottom where it is soft (sand or silt). This downward jet phenomenon was previously described for deltas into the Beaufort Sea, where they were caused by periodic tidal buoyancy of holed ice.

Formed During Storms

Anchor ice may be formed in the shallow intertidal or subtidal during storms in cold weather, when the uppermost layers of the water column are churned up by strong winds or waves. This type of anchor ice can be found primarily in the Arctic, where submerged ice may be observed to completely cover the substrate to depths of up to 2m, with some anchor ice cover observed at more than 4.5m depth.

In the Antarctic

Anchor ice growing on a rope. McMurdo Sound, Antarctica.

Antarctic anchor ice is perhaps one of the most interesting phenomena of ice formation in the marine environment. The general mechanism of its formation is commonly assumed as the following:

- Antarctic surface waters are forced to flow below a large, thick mass of floating ice (ice shelf or glacier tongue) due to tidal motions or normal ocean circulation.

- The surface water melts the underside of the mass of ice, causing a slight freshening of the water that brings the temperature of the water into equilibrium with the in-situ freezing point at depth.

- The water, at its freezing point at depth (slightly lower than the freezing point at the surface due to the pressure effect on the freezing point), is advected from under the floating mass of ice and may rise towards the surface due to a variety of factors.

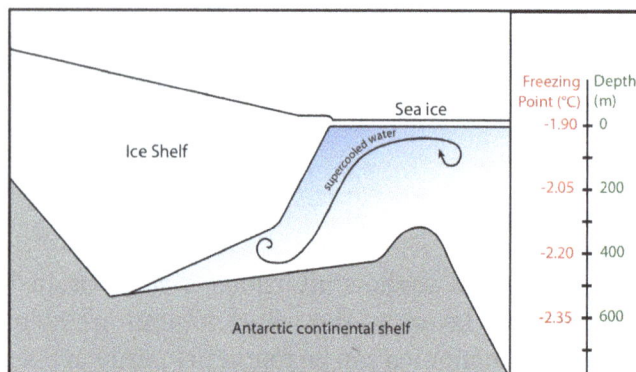

Ice shelves drive the formation of supercooled seawater and anchor ice in the Southern Ocean around Antarctica.

- As the water rises the in-situ freezing point increases, leaving the water slightly supercooled. Supercooling is relieved by the formation of microscopic ice crystals in the water column.

- Ice crystals may coalesce or adhere to submerged objects, including marine organisms, rocks, man-made structures, or other ice formations such as the sea ice, the ice foot, floating glaciers or icebergs.

Anchor ice crystals in the Antarctic are generally in the form of thin, circular platelets of 2–10 cm in diameter. Large masses of irregularly-oriented crystals form anchor ice formations, which may be as large as 4m in diameter when attached to large immovable objects on the sea floor.

Anchor ice that forms on the underside of sea ice is often referred to as platelet or congelation ice, and can be hard to distinguish from that formed due to the cooling of the sea ice cover by cold atmospheric conditions.

Anchor ice is thought to be relatively common in the Antarctic, due to large ice shelves that occupy many areas of the continental coast. Studies and observations of anchor ice formation in McMurdo Sound, Antarctica have shown that the phenomenon regularly causes the formation of ice on the seafloor to depths of approximately 15m, and rarely to depths of approximately 30m.

Biological Effects

A scientist investigating a sponge on the surface of the Western McMurdo Shelf, McMurdo Sound, Antarctica.

Especially in the Antarctic, anchor ice has been implicated in drastic zonation of the subtidal fauna. Many animals are directly affected by the growth of anchor ice, and certain sponges have been shown to readily grow anchor ice and to be damaged by it.

Anchor ice may also grow on animate or inanimate objects and lift them from the sea floor. In the Antarctic this will most likely result in the death of an organism, since during much of the year the ocean is covered by annual sea ice, and the organism is likely to become incorporated into this.

Many organisms have actually been found on the surface of ice shelves in certain places in the Antarctic, likely due to the anchor icing phenomenon:

- Organism accumulates anchor ice as it is bathed in supercooled water.

- Organism becomes positively buoyant due to the accumulation of ice, and it lifted from the sea floor.

- Organism floats to the underside of the ice shelf or sea ice cover where it freezes in.

- Ablation of the surface of the ice cover and additional growth of ice on the underside will result in the organism being transported "through" the ice and "deposited" on the surface, largely intact.

Iceberg

Iceberg in the Arctic with its underside visible

An iceberg or ice mountain is a large piece of freshwater ice that has broken off a glacier or an ice shelf and is floating freely in open water. It may subsequently become frozen into pack ice (one form of sea ice). As it drifts into shallower waters, it may come into contact with the seabed, a process referred to as seabed gouging by ice. Almost 91% of an iceberg is below the surface of the water.

Etymology

The word "iceberg" is a partial loan translation from Dutch *ijsberg*, literally meaning *ice mountain*, cognate to Danish *isbjerg*, German *Eisberg*, Low Saxon *Iesbarg* and Swedish *isberg*.

Overview

Grotto in an iceberg, photographed during the British Antarctic Expedition of 1911–1913, 5 Jan 1911.

Because the density of pure ice is about 920 kg/m³, and that of seawater about 1025 kg/m³, typically only one-tenth of the volume of an iceberg is above water (due to Archimedes's Principle). The shape of the underwater portion can be difficult to judge by looking at the portion above the surface. This has led to the expression "tip of the iceberg", for a problem or difficulty that is only a small manifestation of a larger problem.

Icebergs generally range from 1 to 75 metres (3.3 to 246.1 ft) above sea level and weigh 100,000 to 200,000 metric tons (110,000 to 220,000 short tons). The largest known iceberg in the North Atlantic was 168 metres (551 ft) above sea level, reported by the USCG icebreaker *East Wind* in 1958, making it the height of a 55-story building. These icebergs originate from the glaciers of western Greenland and may have an interior temperature of −15 to −20°C (5 to −4°F).

Icebergs are usually confined by winds and currents to move close to the coast. The largest icebergs recorded have been calved, or broken off, from the Ross Ice Shelf of Antarctica. Iceberg B-15, photographed by satellite in 2000, measured 295 by 37 kilometres (183 by 23 mi), with a surface area of 11,000 square kilometres (4,200 sq mi). The largest iceberg on record was an Antarctic tabular iceberg of over 31,000 square kilometres (12,000 sq mi) [335 by 97 kilometres (208 by 60 mi)] sighted 150 miles (240 km) west of Scott Island, in the South Pacific Ocean, by the USS *Glacier* on November 12, 1956. This iceberg was larger than Belgium.

When a piece of iceberg ice melts, it makes a fizzing sound called "Bergie Seltzer". This sound is made when the water-ice interface reaches compressed air bubbles trapped in the ice. As this happens, each bubble bursts, making a 'popping' sound. The bubbles contain air trapped in snow layers very early in the history of the ice, that eventually got buried to a given depth (up to several kilometers) and pressurized as it transformed into firn then to glacial ice.

Recent Large Icebergs

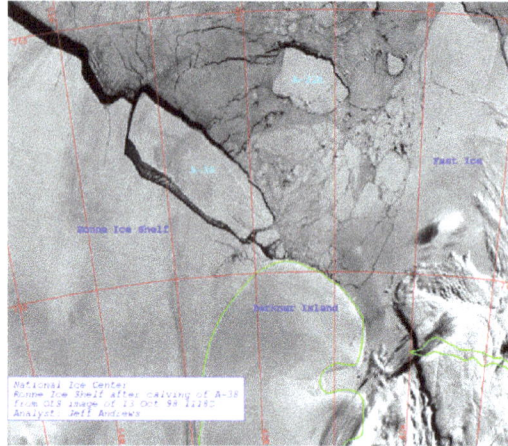

The calving of Iceberg A-38 off Ronne Ice Shelf

- Iceberg B-15 11,000 km² (4,200 sq mi), 2000

- Iceberg A-38, about 6,900 km² (2,700 sq mi), 1998

- Iceberg B-15A, 3,100 km² (1,200 sq mi), broke off 2003

- Iceberg C-19, 5,500 km² (2,100 sq mi), 2002

- Iceberg B-9, 5,390 km² (2,080 sq mi), 1987

- Iceberg B-31, 615 km² (237 sq mi), 2014

- Iceberg D-16, 310 km² (120 sq mi), 2006

- Ice sheet, 260 km² (100 sq mi), broken off of Petermann Glacier in northern Greenland on Aug 5, 2010, considered to be the largest Arctic iceberg since 1962. About a month later, this iceberg split into two pieces upon crashing into Joe Island in the Nares Strait next to Greenland. In June 2011, large fragments of the Petermann Ice Islands were observed off the Labrador coast.

- Iceberg B-17B 140 km² (54 sq mi), 1999, shipping alert issued December 2009.

Shape

In addition to size classification, icebergs can be classified on the basis of their shape. The two basic types of iceberg forms are *tabular* and *non-tabular*. Tabular icebergs have steep sides and a flat top, much like a plateau, with a length-to-height ratio of more than 5:1. This type of iceberg, also known as an *ice island*, can be quite large, as in the case of Pobeda Ice Island. Antarctic icebergs formed by breaking off from an ice shelf, such as the Ross Ice Shelf or Filchner-Ronne Ice Shelf, are typically tabular. The largest icebergs in the world are formed this way.

Tabular iceberg, near Brown Bluff in the Antarctic Sound off Tabarin Peninsula.

Non-tabular iceberg off Elephant Island in the Southern Ocean.

Different shapes of icebergs. 1: Tabular; 2: Wedge; 3: Dome; 4: Drydock; 5: Pinnacled; 6: Blocky.

Non-tabular icebergs have different shapes and include:

- *Dome*: An iceberg with a rounded top.

- *Pinnacle*: An iceberg with one or more spires.

- *Wedge*: An iceberg with a steep edge on one side and a slope on the opposite side.

- *Dry-Dock*: An iceberg that has eroded to form a slot or channel.

- *Blocky*: An iceberg with steep, vertical sides and a flat top. It differs from tabular icebergs in that its shape is more like a block than a flat sheet.

Monitoring

Icebergs are monitored worldwide by the U.S. National Ice Center (NIC), established in 1995, which produces analyses and forecasts of Arctic, Antarctic, Great Lakes and Chesapeake Bay ice conditions. More than 95% of the data used in its sea ice analyses are derived from the remote sensors on polar-orbiting satellites that survey these remote regions of the Earth.

Iceberg A22A in the South Atlantic Ocean

The NIC is the only organization that names and tracks all Antarctic Icebergs. It assigns each iceberg larger than 10 nautical miles (19 km) along at least one axis a name composed of a letter indicating its point of origin and a running number. The letters used are as follows:

A – longitude 0° to 90° W (Bellingshausen Sea, Weddell Sea)

B – longitude 90° W to 180° (Amundsen Sea, Eastern Ross Sea)

C – longitude 90° E to 180° (Western Ross Sea, Wilkes Land)

D – longitude 0° to 90° E (Amery Ice Shelf, Eastern Weddell Sea)

Iceberg B15 calved from the Ross Ice Shelf in 2000 and initially had an area of 11,000 square kilometres (4,200 sq mi). It broke apart in November 2002. The largest remaining piece of it, Iceberg B-15A, with an area of 3,000 square kilometres (1,200 sq mi), was still the largest iceberg on Earth until it ran aground and split into several pieces October 27, 2005, an event that was observed by seismographs both on the iceberg and across Antarctica. It has been hypothesized that this breakup may also have been abetted by ocean swell generated by an Alaskan storm 6 days earlier and 13,500 kilometres (8,400 mi) away.

History

In the 20th century, several scientific bodies were established to study and monitor the icebergs. The International Ice Patrol, formed in 1914 in response to the April 1912 sinking of the *Titanic*, which killed 1,517 of its 2,223 passengers, monitors iceberg dangers near the Grand Banks of Newfoundland and provides the "limits of all known ice" in that vicinity to the maritime community.

The iceberg suspected of sinking the RMS Titanic; a smudge of red paint much like the Titanic's red hull stripe was seen near its base at the waterline.

Technology History

Before the early 1910s there was no system in place to track icebergs to guard ships against collisions, most likely because they weren't considered a serious threat back then, ships have managed to survive even direct crashes. In 1907 *SS Kronprinz Wilhelm*, a German liner, had rammed an iceberg and suffered a crushed bow, but was still able to complete her voyage. The April 1912 sinking of the *Titanic* however changed all that, and created the demand for a system to observe icebergs. For the remainder of the ice season of that year, the United States Navy patrolled the waters and monitored ice flow. In November 1913, the International Conference on the Safety of Life at Sea met in London to devise a more permanent system of observing icebergs. Within three months the participating maritime nations had formed the International Ice Patrol (IIP). The goal of the IIP was to collect data on meteorology and oceanography to measure currents, ice-flow, ocean temperature, and salinity levels. They published their first records in 1921, which allowed for a year-by-year comparison of iceberg movement.

New technologies monitor icebergs. Aerial surveillance of the seas in the early 1930s allowed for the development of charter systems that could accurately detail the ocean currents and iceberg locations. In 1945, experiments tested the effectiveness of radar in detecting icebergs. A decade later, oceanographic monitoring outposts were established for the purpose of collecting data; these outposts continue to serve in environmental study. A computer was first installed on a ship for the purpose of oceanographic monitoring in 1964, which allowed for a faster evaluation of data. By the 1970s, ice-breaking ships were equipped with automatic transmissions of satellite photographs of ice in Antarctica. Systems for optical satellites had been developed but were still limited by weather conditions. In the 1980s, drifting buoys were used in Antarctic waters for oceanographic and climate research. They are equipped with sensors that measure ocean temperature and currents.

An iceberg being pushed by three U.S. Navy ships in McMurdo Sound, Antarctica.

Side looking airborne radar (SLAR) made it possible to acquire images regardless of weather conditions. On November 4, 1995, Canada launched RADARSAT-1. Developed by the Canadian Space Agency, it provides images of Earth for scientific and commercial purposes. This system was the first to use synthetic aperture radar (SAR), which sends microwave energy to the ocean surface and records the reflections to track icebergs. The European Space Agency launched ENVISAT (an observation satellite that orbits the Earth's poles) on March 1, 2002. ENVISAT employs advanced synthetic aperture radar (ASAR) technology, which can detect changes in surface height accurately. The Canadian Space Agency launched RADARSAT-2 in December 2007, which uses SAR and multi-polarization modes and follows the same orbit path as RADARSAT-1.

Blue Iceberg

Blue iceberg discovered during scientific expedition to the coast of Alaska, 2010.

A blue iceberg is visible after the ice from above the water melts, causing the smooth portion of ice from below the water to overturn. The rare blue ice is formed from the compression of pure snow, which then develops into glacial ice.

Icebergs may also appear blue due to light refraction and age. Older icebergs reveal vivid hues of green and blue, resulting from a high concentration of color, microorganisms, and compacted ice. One of the better known blue icebergs rests in the waters off Sermilik fjord near Greenland. It is described as an electric blue iceberg and is known to locals as "blue diamond".

Physics of Light and Color

Blue iceberg seen in the Ilulissat Icefjord, 2015.

White Icebergs

Commonly seen white icebergs generally derive their color from the snow and frost remaining on the surface which results in the uniform reflection of incident light. Young glaciers that have not undergone years of compression, may also appear white. Due to the age of the iceberg, there remains a tremendous amount of air and reflective surfaces. The iceberg easily reflects the sun as white light.

Preferential Light Absorption and Age

Blue icebergs develop from older, deep glaciers which have undergone tremendous pressure experienced for hundreds of years. The process releases and eliminates air that was originally caught in the ice by falling snow. Therefore, icebergs that have been formed from older glaciers have little internal air or reflective surfaces. When long wavelength light (i.e. red) from the sun hits the iceberg, it is absorbed, rather than reflected. The light transmitted or refracted through the ice returns as blue or blue-green. Older glaciers also reflect incident light preferentially at the short wavelength end of the spectrum (i.e. blue) due to Rayleigh scattering, much in the same way that makes the sky blue.

Color Spectrum and Water

Light is absorbed and reflected in water. Visible white light is made up of a spectrum of colors from the rainbow, ranging from red to violet. As the light travels through the water, the waves of light from the red end of the spectrum dissipate (i.e. are absorbed), while those from the blue end, become more prominent.

SCUBA divers have direct experience of these effects. Above the water, all the colors remain visible. As the diver swims deeper under water, the colors begin to disappear, starting with red. At the same time, the other colors are absorbed by the water and cannot be seen. At an approximate depth of 30 feet (9.1 m), red is no longer visible to the naked eye. At 75 feet (23 m), yellow looks greenish-blue, because the water has absorbed the yellow light. Finally, all that remains visible to the naked eye, appears as a mutation of blue or green, while the water above the surface filters out the sunlight. As the diver swims deeper into the ocean, he finds that the blue colors start to disappear, to the point where the underwater world deep below the surface, becomes completely black, devoid of any color at all.

RMS Titanic

Since 1912, reports made by witnesses of the RMS *Titanic* tragedy have stated that the ship hit a blue iceberg. Following the sinking and subsequent discovery of the *Titanic*, scientific research and forensic analysis have reconstructed the tragedy to ascertain the reliability of the statements made by the survivors. Reports released in the last decade of the 20th century have shown that a blue iceberg in the north Atlantic would have been easily detected. Alternative theories suggest that pack ice, rather than a blue iceberg, was responsible for sinking the ship.

Iceberg B-9

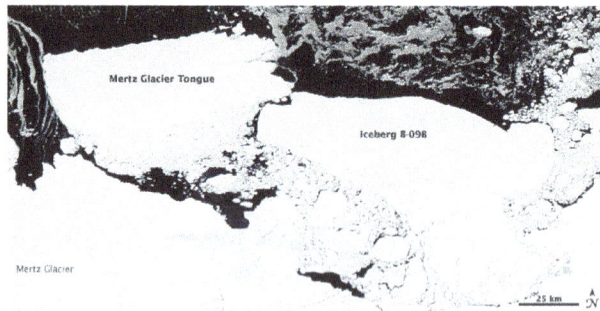

Iceberg B-9B colliding with the Mertz Glacier Tongue calving the Mertz iceberg, 20 February 2010.

Iceberg B-9 was an iceberg that calved in 1987. The iceberg measured 154 kilometres (96 mi) long and 35 kilometres (22 mi) wide with a total area of 5,390 square kilometres (2,080 sq mi). It is one of the longest icebergs ever recorded. The calving took place immediately east of the calving site of Iceberg B-15 and carried away Little America V.

Starting in October 1987, Iceberg B-9 drifted for 22 months and covered 2,000 kilometres (1,200 mi) on its journey. Initially, B-9 moved northwest for seven months before being drawn southward by a subsurface current that eventually led to its colliding with the Ross Ice Shelf in August 1988. It then made a 100-kilometre (62 mi) radius gyre before continuing its northwest drift. B-9 moved at an average speed of 2.5 kilometres (1.6 mi) per day over the continental shelf, as measured by NOAA-10 and DMSP satellite positions, and the ARGOS data buoy positions. In early August 1989, B-9 broke into three large pieces north of Cape Adare. These pieces were B-9A, 56 by 35 kilometres (35 mi × 22 mi), B-9B, 100 by 35 kilometres (62 mi × 22 mi), and B-9C, 28 by 13 kilometres (17.4 mi × 8.1 mi).

B-9B drifted toward the Mertz Glacier on the George V Coast, where it came to rest next to the glacier and remained there for eighteen years. On February 12 or 13th 2010, Iceberg B-9B collided with the giant floating Mertz Glacier tongue and shaved off a new iceberg that measured 78 kilometres (48 mi) long and 39 kilometres (24 mi) wide. The two icebergs then began drifting together about 100–150 kilometres (62–93 mi) off the eastern coast of Antarctica.

By December 2011, Iceberg B-9B had made its way into Commonwealth Bay and had broken up into three major pieces, parts of which were frozen to the seabed. The huge iceberg prevented three tourist ships from reaching Antarctica to mark the centenary of the polar voyage of Australian explorer Douglas Mawson, who landed at Cape Denison on January 8, 1912 and constructed a complex of huts that remain standing to this day. The three tourist ships attempted to reach the cape but had to turn back due to unusually harsh conditions caused by B-9B's position in the bay. A spokeswoman from the Australian government's Antarctic division observed, "There [are] unusual ice conditions… affecting all the tourist ships that are going down there because the tourist ships don't have ice-breaking capabilities, and they also don't have choppers, so their ability to get anywhere near the Mawson's huts area is basically stopped." Iceberg B-9B could remain in Commonwealth Bay for the next decade.

A report in the press announced that since 2011, an estimated 150,000 Adélie penguins living in Antarctica have died after B-9B became grounded near their colony at Cape Denison in Commonwealth Bay. The iceberg effectively blocked the penguins' access to the ocean and their primary source of food; they must now trek 60 kilometres (37 mi) to the sea. According to the Climate Change Research Center at Australia's University of New South Wales, the colony of 160,000 penguins has been reduced to only 10,000. Scientists predict that this colony will be gone in 20 years unless the sea ice breaks up or B-9B is dislodged. However, subsequent release from experts from the Scientific Committee on Antarctic Research (SCAR) and elsewhere in the world corrected the over-statement made by the journalists. The missing penguins did not die because of the presence of the iceberg but most probably left to establish new colonies in more suitable places, as the original publication suggests.

Ice Cap

An ice cap is an ice mass that covers less than 50,000 km² of land area (usually covering a highland area). Larger ice masses covering more than 50,000 km² are termed ice sheets

Vatnajökull, Iceland

Ice caps are not constrained by topographical features (i.e., they will lie over the top of mountains). By contrast, ice masses of similar size that *are* constrained by topographical features are known as ice fields. The *dome* of an ice cap is usually centred on the highest point of a massif. Ice flows away from this high point (the ice divide) towards the ice cap's periphery.

Ice caps have significant effects on the geomorphology of the area they occupy. Plastic moulding, gouging and other glacial erosional features become present upon the glacier's retreat. Many lakes, such as the Great Lakes in North America, as well as numerous valleys have been formed by glacial action over hundreds of thousands of years.

On Earth, there are about 30 million km³ of total ice mass. The average temperature of an ice mass ranges between −20 °C and −30 °C. The core of an ice cap exhibits a constant temperature that ranges between −15 °C and −20 °C.

A high-latitude region covered in ice, though strictly not an ice cap (since they exceed the maximum area specified in the definition above), are called polar ice caps; the usage of this designation is widespread in the mass media and arguably recognized by experts. Vatnajökull is an example of an ice cap in Iceland.

Ice Sheet

An ice sheet is a mass of glacier ice that covers surrounding terrain and is greater than 50,000 km² (19,000 sq mi), thus also known as continental glacier. The only current ice sheets are in Antarctica and Greenland; during the last glacial period at Last Glacial Maximum (LGM) the Laurentide ice sheet covered much of North America, the Weichselian ice sheet covered northern Europe and the Patagonian Ice Sheet covered southern South America.

A satellite composite image of Antarctica

Ice sheets are bigger than ice shelves or alpine glaciers. Masses of ice covering less than 50,000 km² are termed an ice cap. An ice cap will typically feed a series of glaciers around its periphery.

Aerial view of the ice sheet on Greenland's east coast

Although the surface is cold, the base of an ice sheet is generally warmer due to geothermal heat. In places, melting occurs and the melt-water lubricates the ice sheet so that it flows more rapidly. This process produces fast-flowing channels in the ice sheet — these are ice streams.

The present-day polar ice sheets are relatively young in geological terms. The Antarctic Ice Sheet first formed as a small ice cap (maybe several) in the early Oligocene, but retreating and advancing many times until the Pliocene, when it came to occupy almost all of Antarctica. The Greenland ice sheet did not develop at all until the late Pliocene, but apparently developed *very rapidly* with the first continental glaciation. This had the unusual effect of allowing fossils of plants that once grew on present-day Greenland to be much better preserved than with the slowly forming Antarctic ice sheet.

Antarctic Ice Sheet

The Antarctic ice sheet is the largest single mass of ice on Earth. It covers an area of almost 14 million km² and contains 30 million km³ of ice. Around 90% of the Earth's ice mass is in Antarctica, which, if melted, would cause sea levels to rise by 58 meters. The continent-wide average surface temperature trend of Antarctica is positive and significant at >0.05 °C/decade since 1957.

The Antarctic ice sheet is divided by the Transantarctic Mountains into two unequal sections called the East Antarctic ice sheet (EAIS) and the smaller West Antarctic Ice Sheet (WAIS). The EAIS rests on a major land mass but the bed of the WAIS is, in places, more than 2,500 metres below sea level. It would be seabed if the ice sheet were not there. The WAIS is classified as a marine-based ice sheet, meaning that its bed lies below sea level and its edges flow into floating ice shelves. The WAIS is bounded by the Ross Ice Shelf, the Ronne Ice Shelf, and outlet glaciers that drain into the Amundsen Sea.

Greenland Ice Sheet

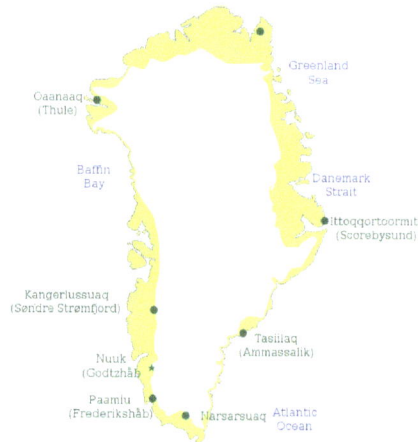

Map of Greenland

The Greenland ice sheet occupies about 82% of the surface of Greenland, and if melted would cause sea levels to rise by 7.2 metres. Estimated changes in the mass of Greenland's ice sheet suggest it is melting at a rate of about 239 cubic kilometres (57.3 cubic miles) per year. These measurements came from NASA's Gravity Recovery and Climate Experiment (GRACE) satellite, launched in 2002, as reported by BBC News in August 2006.

Ice Sheet Dynamics

Ice movement is dominated by the motion of glaciers, whose activity is determined by a number of processes. Their motion is the result of cyclic surges interspersed with longer periods of inactivity, on both hourly and centennial time scales.

Predicted Effects of Global Warming

The Greenland, and possibly the Antarctic, ice sheets have been losing mass recently, because losses by ablation including outlet glaciers exceed accumulation of snowfall.

According to the Intergovernmental Panel on Climate Change (IPCC), loss of Antarctic and Greenland ice sheet mass contributed, respectively, about 0.21 ± 0.35 and 0.21 ± 0.07 mm/year to sea level rise between 1993 and 2003.

The IPCC projects that ice mass loss from melting of the Greenland ice sheet will continue to outpace accumulation of snowfall. Accumulation of snowfall on the Antarctic ice sheet is projected to outpace losses from melting. However, in the words of the IPCC, *"Dynamical processes related to ice flow not included in current models but suggested by recent observations could increase the vulnerability of the ice sheets to warming, increasing future sea level rise. Understanding of these processes is limited and there is no consensus on their magnitude."* More research work is therefore required to improve the reliability of predictions of ice-sheet response on global warming.

The effects on ice sheets due to increasing temperature may accelerate, but as documented by the IPCC the effects are not easily projected accurately and in the case of the Antarctic, may trigger an accumulation of additional ice mass. If an ice sheet were ablated down to bare ground, less light from the sun would be reflected back into space and more would be absorbed by the land. The Greenland Ice Sheet covers 84% of the island and the Antarctic Ice Sheet covers approximately 98% of the continent. Due to the significant thickness of these ice sheets, global warming analysis typically focuses on the loss of ice mass from the ice sheets increasing sea level rise, and not on a reduction in the surface area of the ice sheets.

Ice-sheet Dynamics

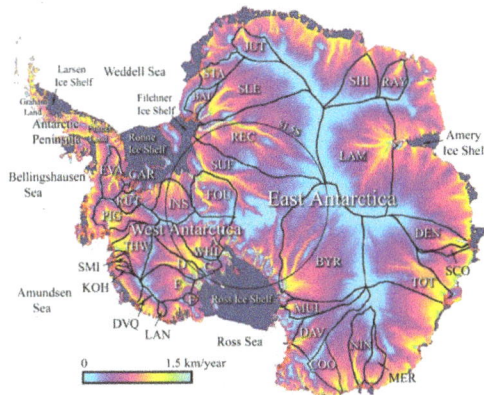

Glacial flow rate in the Antarctic ice sheet

Ice sheet dynamics describe the motion within large bodies of ice, such those currently on Greenland and Antarctica. Ice motion is dominated by the movement of glaciers, whose gravity-driven activity is controlled by two main variable factors: the temperature and strength of their bases. A number of processes alter these two factors, resulting in cyclic surges of activity interspersed with longer periods of inactivity, on both hourly and centennial time scales. Ice-sheet dynamics are of interest in modelling future sea level rise.

Flow Dynamics

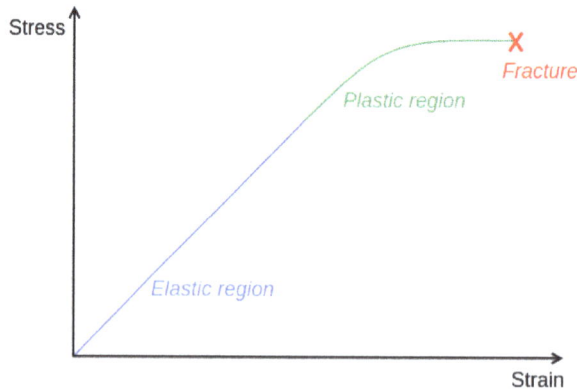

The stress–strain relationship of plastic flow (teal section): a small increase in stress creates an exponentially greater increase in strain, which equates to deformation speed.

The main cause of flow within glaciers can be attributed to an increase in the surface slope, brought upon by an imbalance between the amounts of accumulation vs. ablation. This imbalance increases the shear stress on a glacier until it begins to flow. The flow velocity and deformation will increase as the equilibrium line between these two processes is approached, but are also affected by the slope of the ice, the ice thickness and temperature.

When the amount of strain (deformation) is proportional to the stress being applied, ice will act as an elastic solid. Ice will not flow until it has reached a thickness of 30 meters (98 ft), but after 50 meters (164 ft), small amounts of stress can result in a large amount of strain, causing the deformation to become a plastic flow rather than elastic. At this point the glacier will begin to deform under its own weight and flow across the landscape. According to the Glen–Nye Flow law, the relationship between stress and strain, and thus the rate of internal flow, can be modeled as follows:

$$\Sigma = k\tau^n,$$

where:

Σ = shear strain (flow) rate

τ = stress

n = a constant between 2–4 (typically 3 for most glaciers) that increases with lower temperature

k = a temperature-dependent constant

The lowest velocities are near the base of the glacier and along valley sides where friction acts against flow, causing the most deformation. Velocity increases inward toward the center line and upward, as the amount of deformation decreases. The highest flow velocities are found at the surface, representing the sum of the velocities of all the layers below.

Glaciers may also move by basal sliding, where the base of the glacier is lubricated by meltwater, allowing the glacier to slide over the terrain on which it sits. Meltwater may be produced by pressure-induced melting, friction or geothermal heat. The more variable the amount of melting at surface of the glacier, the faster the ice will flow.

The top 50 meters of the glacier form the fracture zone, where ice moves as a single unit. Cracks form as the glacier moves over irregular terrain, which may penetrate the full depth of the fracture zone.

Subglacial Processes

A cross-section through a glacier. The base of the glacier is more transparent as a result of melting.

Most of the important processes controlling glacial motion occur in the ice-bed contact—even though it is only a few meters thick. Glaciers will move by sliding when the basal shear stress drops below the shear resulting from the glacier's weight.

$$\tau_D = \rho g h \sin \alpha$$

where τ_D is the driving stress, and α the ice surface slope in radians.

τ_B is the basal shear stress, a function of bed temperature and softness.

τ_F, the shear stress, is the lower of τ_B and τ_D. It controls the rate of plastic flow, as per the figure.

For a given glacier, the two variables are τ_D, which varies with h, the depth of the glacier, and τ_B, the basal shear stress.

Basal Shear Stress

The basal shear stress is a function of three factors: the bed's temperature, roughness and softness.

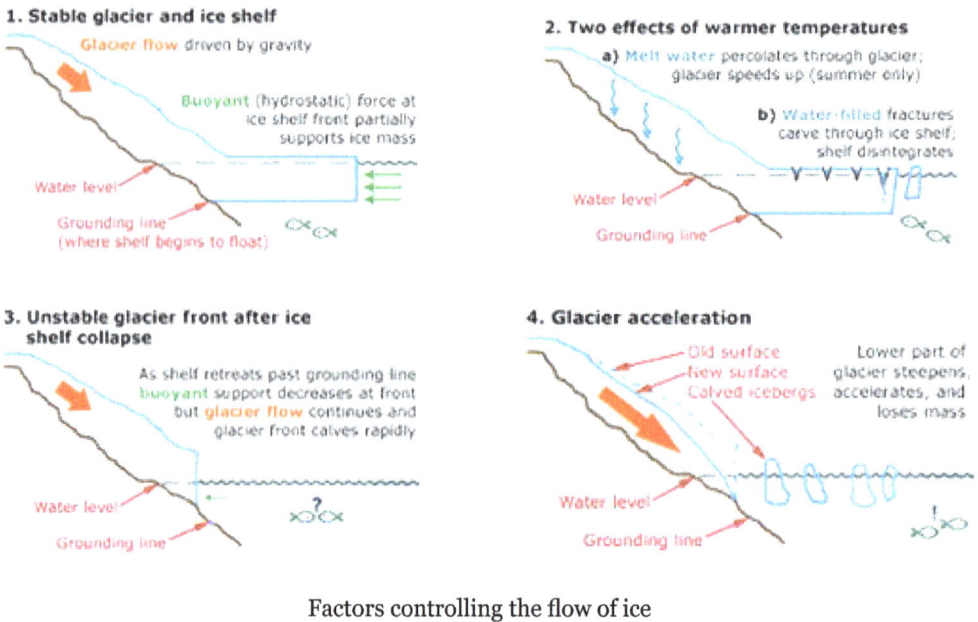

1. Stable glacier and ice shelf

Glacier flow driven by gravity

Buoyant (hydrostatic) force at ice shelf front partially supports ice mass

Water level

Grounding line (where shelf begins to float)

2. Two effects of warmer temperatures

a) Melt water percolates through glacier; glacier speeds up (summer only)

b) Water-filled fractures carve through ice shelf; shelf disintegrates

Water level

Grounding line

3. Unstable glacier front after ice shelf collapse

As shelf retreats past grounding line buoyant support decreases at front but glacier flow continues and glacier front calves rapidly

Water level

Grounding line

4. Glacier acceleration

Old surface
New surface
Calved icebergs

Lower part of glacier steepens, accelerates, and loses mass

Water level

Grounding line

Factors controlling the flow of ice

Whether a bed is hard or soft depends on the porosity and pore pressure; higher porosity decreases the sediment strength (thus increases the shear stress τ_B). If the sediment strength falls far below τ_D, movement of the glacier will be accommodated by motion in the sediments, as opposed to sliding. Porosity may vary through a range of methods.

- Movement of the overlying glacier may cause the bed to undergo dilatancy; the resulting shape change reorganises blocks. This reorganises closely packed blocks (a little like neatly folded, tightly packed clothes in a suitcase) into a messy jumble (just as clothes never fit back in when thrown in in a disordered fashion). This increases the porosity. Unless water is added, this will necessarily reduce the pore pressure (as the pore fluids have more space to occupy).

- Pressure may cause compaction and consolidation of underlying sediments. Since water is relatively incompressible, this is easier when the pore space is filled with vapour; any water must be removed to permit compression. In soils, this is an irreversible process.

- Sediment degradation by abrasion and fracture decreases the size of particles, which tends to decrease pore space, although the motion of the particles may disorder the sediment, with the opposite effect. These processes also generate heat, whose importance will be discussed later.

Bed softness may vary in space or time, and changes dramatically from glacier to glacier. An important factor is the underlying geology; glacial speeds tend to differ more when they change bedrock than when the gradient changes.

As well as affecting the sediment stress, fluid pressure (p_w) can affect the friction between the glacier and the bed. High fluid pressure provides a buoyancy force upwards on the glacier, reducing the friction at its base. The fluid pressure is compared to the ice overburden pressure, p_i, given by $\rho g h$. Under fast-flowing ice streams, these two pressures will be approximately equal, with an effective pressure ($p_i - p_w$) of 30 kPa; i.e. all of the weight of the ice is supported by the underlying water, and the glacier is afloat.

Basal Melt

A number of factors can affect bed temperature, which is intimately associated with basal meltwater. The melting point of water decreases under pressure, meaning that water melts at a lower temperature under thicker glaciers. This acts as a "double whammy", because thicker glaciers have a lower heat conductance, meaning that the basal temperature is also likely to be higher.

Bed temperature tends to vary in a cyclic fashion. A cool bed has a high strength, reducing the speed of the glacier. This increases the rate of accumulation, since newly fallen snow is not transported away. Consequently, the glacier thickens, with three consequences: firstly, the bed is better insulated, allowing greater retention of geothermal heat. Secondly, the increased pressure can facilitate melting. Most importantly, τ_D is increased. These factors will combine to accelerate the glacier. As friction increases with the square of velocity, faster motion will greatly increase frictional heating, with ensuing melting – which causes a positive feedback, increasing ice speed to a faster flow rate still: west Antarctic glaciers are known to reach velocities of up to a kilometre per year. Eventually, the ice will be surging fast enough that it begins to thin, as accumulation cannot keep up with the transport. This thinning will increase the conductive heat loss, slowing the glacier and causing freezing. This freezing will slow the glacier further, often until it is stationary, whence the cycle can begin again.

Supraglacial lakes represent another possible supply of liquid water to the base of glaciers, so they can play an important role in accelerating glacial motion. Lakes of a diameter greater than ~300 m are capable of creating a fluid-filled crevasse to the glacier/bed interface. When these crevasses form, the entirety of the lake's (relatively warm) contents can reach the base of the glacier in as little as 2–18 hours – lubricating the bed and causing the glacier to surge. Water that reaches the bed of a glacier may freeze there, increasing the thickness of the glacier by pushing it up from below.

Finally, bed roughness can act to slow glacial motion. The roughness of the bed is a measure of how many boulders and obstacles protrude into the overlying ice. Ice flows around these obstacles by melting under the high pressure on their lee sides; the resultant meltwater is then forced down a steep pressure gradient into the cavity arising in their stoss, where it re-freezes. Cavitation on the stoss side increases this pressure gradient, which assists flow.

Erosional Effects

Because ice can flow faster where it is thicker, the rate of glacier-induced erosion is directly proportional to the thickness of overlying ice. Consequently, pre-glacial low hollows will be deepened and pre-existing topography will be amplified by glacial action, while nunataks, which protrude above ice sheets, barely erode at all – erosion has been estimated as 5 m per 1.2 million years. This explains, for example, the deep profile of fjords, which can reach a kilometer in depth as ice is topographically steered into them. Being the principal conduits for draining ice sheets, fjords' extension inland increases the rate of ice sheet thinning. It also makes the ice sheets more sensitive to changes in climate and the ocean.

Differential erosion enhances relief, as clear in this incredibly steep-sided Norwegian fjord.

Pipe and Sheet Flow

The flow of water under the glacial surface can have a large effect on the motion of the glacier itself. Subglacial lakes contain significant amounts of water, which can move fast: cubic kilometres can be transported between lakes over the course of a couple of years.

This motion is thought to occur in two main modes: pipe flow involves liquid water moving through pipe-like conduits, like a sub-glacial river; sheet flow involves motion of water in a thin layer. A switch between the two flow conditions may be associated with surging behaviour. Indeed, the loss of sub-glacial water supply has been linked with the shut-down of ice movement in the Kamb ice stream. The subglacial motion of water is expressed in the surface topography of ice sheets, which slump down into vacated subglacial lakes.

Boundary Conditions

The interface between an ice stream and the ocean is a significant control of the rate of flow.

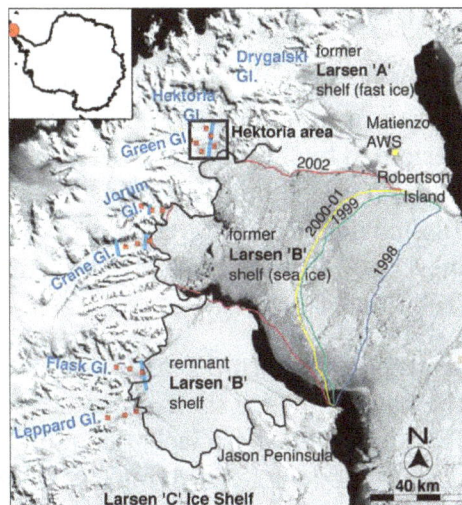

The collapse of the Larsen B ice shelf had profound effects on the velocities of its feeder glaciers.

Ice shelves – thick layers of ice floating on the sea – can stabilise the glaciers that feed them. These tend to have accumulation on their tops, may experience melting on their bases, and calve icebergs at their periphery. The catastrophic collapse of the Larsen B ice shelf in the space of three weeks during February 2002 yielded some unexpected observations. The glaciers that had fed the ice sheet increased substantially in velocity. This cannot have been due to seasonal variability, as glaciers flowing into the remnants of the ice shelf (Flask, Leppard) did not accelerate.

Ice shelves exert a dominant control in Antarctica, but are less important in Greenland, where the ice sheet meets the sea in fjords. Here, melting is the dominant ice removal process, resulting in predominant mass loss occurring towards the edges of the ice sheet, where icebergs are calved in the fjords and surface meltwater runs into the ocean.

Tidal effects are also important; the influence of a 1 m tidal oscillation can be felt as much as 100 km from the sea. On an hour-to-hour basis, surges of ice motion can be modulated by tidal activity. During larger spring tides, an ice stream will remain almost stationary for hours at a time, before a surge of around a foot in under an hour, just after the peak high tide; a stationary period then takes hold until another surge towards the middle or end of the falling tide. At neap tides, this interaction is less pronounced, without tides surges would occur more randomly, approximately every 12 hours.

Ice shelves are also sensitive to basal melting. In Antarctica, this is driven by heat fed to the shelf by the circumpolar deep water current, which is 3 °C above the ice's melting point.

As well as heat, the sea can also exchange salt with the oceans. The effect of latent heat, resulting from melting of ice or freezing of sea water, also has a role to play. The effects of these, and variability in snowfall and base sea level combined, account for around 80 mm a^{-1} variability in ice shelf thickness.

Long-term Changes

Over long time scales, ice sheet mass balance is governed by the amount of sunlight reaching the earth. This variation in sunlight reaching the earth, or insolation, over geologic time is in turn determined by the angle of the earth to the sun and shape of the Earth's orbit, as it is pulled on by neighboring planets; these variations occur in predictable patterns called Milankovitch cycles. Milankovitch cycles dominate climate on the glacial–interglacial timescale, but there exist variations in ice sheet extent that are not linked directly with insolation.

For instance, during at least the last 100,000 years, portions of the ice sheet covering much of North America, the Laurentide ice sheet broke apart sending large flotillas of icebergs into the North Atlantic. When these icebergs melted they dropped the boulders and other continental rocks they carried, leaving layers known as ice rafted debris. These so-called Heinrich events, named after their discoverer Hartmut Heinrich, appear to have a 7,000–10,000-year periodicity, and occur during cold periods within the last interglacial.

Internal ice sheet "binge-purge" cycles may be responsible for the observed effects, where the ice builds to unstable levels, then a portion of the ice sheet collapses. External factors might also play a role in forcing ice sheets. Dansgaard–Oeschger events are abrupt warmings of the northern hemisphere occurring over the space of perhaps 40 years. While these D–O events occur directly after each Heinrich event, they also occur more frequently – around every 1500 years; from this evidence, paleoclimatologists surmise that the same forcings may drive both Heinrich and D–O events.

Hemispheric asynchrony in ice sheet behavior has been observed by linking short-term spikes of methane in Greenland ice cores and Antarctic ice cores. During Dansgaard–Oeschger events, the northern hemisphere warmed considerably, dramatically increasing the release of methane from wetlands, that were otherwise tundra during glacial times. This methane quickly distributes evenly across the globe, becoming incorporated in Antarctic and Greenland ice. With this tie, paleoclimatologists have been able to say that the ice sheets on Greenland only began to warm after the Antarctic ice sheet had been warming for several thousand years. Why this pattern occurs is still open for debate.

Effects of Climate Change on Ice Sheet Dynamics

The implications of the current climate change on ice sheets are difficult to ascertain. It is clear that increasing temperatures are resulting in reduced ice volumes globally. (Due to increased precipitation, the mass of parts of the Antarctic ice sheet may currently be increasing, but the total mass balance is unclear.)

Since the surging nature of ice sheet motion is a relatively recent discovery, and is still a long way from being entirely understood, no models have yet made a comprehensive evaluation of the effects of climate change. However, it is clear that climate change will act to destabilise ice sheets by a number of mechanisms.

Rising sea levels will reduce the stability of ice shelves, which have a key role in reducing glacial motion. Some Antarctic ice shelves are currently thinning by tens of metres per year, and the collapse of the Larsen B shelf was preceded by thinning of just 1 metre per year. Further, increased ocean temperatures of 1 °C may lead to up to 10 metres per year of basal melting. Ice shelves are always stable under mean annual temperatures of −9 °C, but never stable above −5 °C; this places regional warming of 1.5 °C, as preceded the collapse of Larsen B, in context.

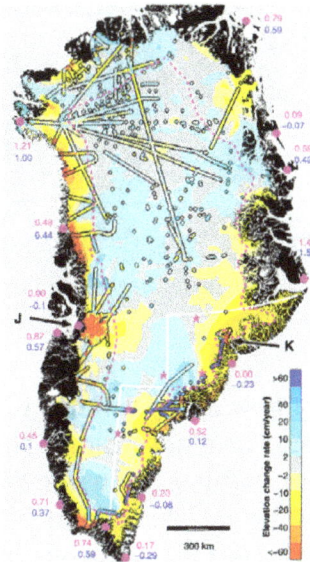

Rates of ice-sheet thinning in Greenland

Increasing global air temperatures take around 10,000 years to directly propagate through the ice before they influence bed temperatures, but may have an effect through increased surfacal melting, producing more supraglacial lakes, which may feed warm water to glacial bases and facilitate glacial motion. In areas of increased precipitation, such as Antarctica, the addition of mass will increase rate of glacial motion, hence the turnover in the ice sheet. Observations, while currently limited in scope, do agree with these predictions of an increasing rate of ice loss from both Greenland and Antarctica. A possible positive feedback may result from shrinking ice caps, in volcanically active Iceland at least. Isostatic rebound may lead to increased volcanic activity, causing basal warming – and, through CO_2 release, further climate change.

Ice Dam

An ice dam occurs when water builds up behind a blockage of ice. Ice dams form either when glacier blocks a river and forms a lake or when ice chunks in a river are blocked by something and build up to form a dam, often called an ice jam. Glacial ice dams have historically resulted in massive outburst floods. River ice jams can cause flooding upstream during the jam, flooding downstream when the jam releases, and damage from the ice itself on structures and ships in or near the river. *Ice jams* on a lake or ocean occur during the spring break-up if wind driven ice piles up along a shoreline.

Niagara falls ice dams block the falling water

Glacial Ice Dams

The movement of a glacier may flow down a valley to a confluence where the other branch carries an unfrozen river. The glacier blocks the river, which backs up into a proglacial lake, which eventually overflows or undermines the ice dam, suddenly releasing the impounded water in a glacial lake outburst flood also known by its Icelandic name a jökulhlaup. Some of the largest glacial floods in North American history were from Lake Agassiz. In modern times, the Hubbard Glacier regularly blocks the mouth of Russell Fjord at 60° north on the coast of Alaska.

A similar event takes place after irregular periods in the Perito Moreno Glacier, located in Patagonia. Roughly every four years the glacier forms an ice dam against the rocky coast, causing the waters of the Lago Argentino to rise. When the water pressure is too high, then the giant bridge collapses in what has become a major tourist attraction. This sequence occurred last on March 4, 2012, the previous having taken place four years before, in July 2008.

About 13,000 years ago in North America, the Cordilleran ice sheet crept southward into the Idaho Panhandle, forming a large ice dam that blocked the mouth of the Clark Fork River, creating a massive lake 2,000 feet (600 m) deep and containing more than 500 cubic miles (2,000 km³) of water. Finally this Glacial Lake Missoula burst through the ice dam and exploded downstream, flowing at a rate 10 times the combined flow of all the rivers of the world. Because such ice dams can re-form, these Missoula Floods happened at least 59 times, carving Dry Falls below Grand Coulee.

River Ice Jams

An ice blockage on a river is more often called an *ice jam* but sometimes an *ice dam*. An ice jam is a dam on a river formed by blocks of ice. Defined by the International Association of Hydraulic Research (IAHR) Working Group on River Ice Hydraulics an ice jam is a "a stationary accumulation of fragmented ice or frazil that restricts flow." on a river or stream. This definition includes what some scientists call an ice dam as a "bottom accumulation of anchor ice".

The town of Eagle, Alaska, is inundated by flood water and ice flows after an ice jam formed on the Yukon River downstream.

Ice jam floods are less predictable and potentially more destructive than open-water flooding and can produce much deeper and faster flooding. Ice jam floods also may occur during freezing weather, and may leave large pieces of ice behind, but they are much more localized than open-water floods. Ice jams also damage an economy by causing river-side industrial facilities such as hydro-electric generating stations to shut down and to interfere with ship transport. The United States averages 125 million dollars in losses to ice jams per year.

Ice jams on rivers usually occur in the springtime as the river ice begins to *break up*, but may also occur in early winter during *freeze-up*. The break-up process is described in three phases: pre-break-up, break-up and final drive. *Pre-break-up* usually begins with increased springtime river flow, water level, and temperatures fracturing the river ice and separating it from the shore. Changes in river height from dam releases may also affect the pre-break-up. During the *break-up*, the ice in areas of rapids is carried downstream as an ice floe and may jam on still frozen sections of ice on calm water or against structures in the river such as the Honeymoon Bridge, destroyed in 1938 by an ice jam. Smaller jams may dislodge, flow downstream and form a larger jam. During the *final drive*, a large jam will dislodge and take out the remaining jams, clearing the river of ice in a matter of hours. Ice jams usually occur in spring, but they can happen as winter sets in when the downstream part becomes frozen first. Freeze-up jams may be larger because the ice is stronger and temperatures are continuing to cool unlike a spring break-up when the environment is warming, but are less likely to suddenly release water.

Three types of natural ice jams can occur:

1. a *surface jam*, a single layer of ice in a floe on calm water;

2. a *narrow-channel* or *wide-channel jam*; and

3. a *hanging jam*, the accumulation of river ice at slow current areas which only occur during freeze-up. Ice jams also occur at sharp bends in the river, at man-made objects such as bridge piers, and at confluences.

Northerly flowing rivers tend to have more ice jams because the upper, more southerly reaches thaw first and the ice gets carried downstream into the still-frozen northerly part. Three physical hazards of ice jams are 1.) The ice floe can form an ice dam and flood the areas upstream of the jam. This occurred during the 2009 Red River Flood and the 2009 Alaska floods. 2.) After the ice dam breaks apart, the sudden surge of water that breaks through the dam can then flood areas downstream of the jam. 3.) The ice buildup and final drive may damage structures in or near the river and boats in the river.

Ice jams may scour the river bed, causing damage or benefit to wildlife habitats and possibly damage to structures in the river.

Early warnings of an ice jam include using trained observers to monitor break-up conditions and ice motion detectors.

The prevention of ice jams may be accomplished by

1. weakening the ice before the break-up by cutting or drilling holes in the ice;

2. weakening the ice by dusting it with a dark colored sand; or

3. controlling the timing of the break-up using ice breakers, towboats, hovercraft, or amphibious excavators. However, the movement of migratory fish is known to be related to freeze-up and break-up, so affecting ice break-up may affect fish migration.

Where floods threaten human habitation, the blockage may be artificially cleared. Ice blasting using dynamite may be used, except in urban areas, as well as other mechanical means such as excavation equipment, or permanent measures such as ice control structures and flood control.

Ice jams on a lake or ocean occur during the spring break-up if wind-driven ice piles up along a shoreline.

Ice Bridge

An ice bridge is a frozen natural structure formed over seas, bays, rivers or lake surfaces. They facilitate migration of animals or people over a water body that was previously uncrossable by terrestrial animals, including humans. The most significant ice bridges are formed by glaciation, spanning distances of many miles over sometimes relatively deep water bodies.

The ice bridge on the St. Lawrence river between Québec and Lévis, Canada, in 1892.

An example of such a major ice bridge was that connecting the island of Öland with mainland Sweden approximately 9000 BC. This bridge reached its maximum utility when the glacier was in retreat, forming a low-lying frozen bridge. The Öland ice bridge allowed the first human migration to the island of Öland, which is most readily documented by archaeological studies of the Alby People.

In Jules Verne's 1873 novel *The Fur Country*, a group of fur trappers establishes a fort on what they think is stable ground, only to find later on that is merely an iceberg temporarily attached by an ice bridge to the mainland.

References

- Warren, S. G.; Roesler, C. S.; Morgan, V. I.; Brandt, R. E.; Goodwin, I. D.; and Allison, I. (1993). "Green icebergs formed by freezing of organic-rich seawater to the base of Antarctic ice shelves" Journal of Geophysical Research Oceans, 98, Volume: 98, Issue: C4, William Byrd Press for Johns Hopkins Press, pp. 6921-6928, 1993

- McCarty, Jennifer Hooper; Foecke, Tim. What Really Sank the Titanic: New Forensic Discoveries, Kensington Publishing Corporation, page 67, 2009. ISBN 978-0-8065-2896-0

- "A World of Ice {in Pictures} | Ice Stories: Dispatches From Polar Scientists". Icestories.exploratorium.edu. 2008-02-23. Retrieved 2011-07-18

- Schoof, C. (2010). "Ice-sheet acceleration driven by melt supply variability". Nature. 468 (7325): 803–806. Bibcode:2010Natur.468..803S. PMID 21150994. doi:10.1038/nature09618

- Sherratt, Thomas N.; and Wilkinson, David M. Big questions in ecology and evolution, Oxford University Press US, page 172, 2009. ISBN 978-0-19-954861-3

- Flowers, Gwenn E.; Shawn J. Marshall; Helgi Björnsson; Garry K. C. Clarke (2005). "Sensitivity of Vatnajökull ice cap hydrology and dynamics to climate warming over the next 2 centuries" (PDF). Journal of Geophysical Research. 110: F02011. Bibcode:2005JGRF..11002011F. doi:10.1029/2004JF000200. Retrieved 2007-05-31

- Bennett, Matthew; Neil Glasser (1996). Glacial Geology: Ice Sheets and Landforms. Chichester, England: John Wiley and Sons Ltd. ISBN 0-471-96345-3

- Keys, Harry; Jacobs, S.S.; Barnett, Don (11 June 1990). "The calving and drift of iceberg B-9 in the Ross Sea, Antarctica". Antarctic Science. 2 (3): 243–257. doi:10.1017/s0954102090000335. Retrieved 23 February 2014

- Bindschadler, A.; King, A.; Alley, B.; Anandakrishnan, S.; Padman, L. (Aug 2003). "Tidally Controlled Stick-Slip Discharge of a West Antarctic Ice". Science. 301 (5636): 1087–1089. ISSN 0036-8075. PMID 12934005. doi:10.1126/science.1087231

- Greve, R.; Blatter, H. (2009). Dynamics of Ice Sheets and Glaciers. Springer. ISBN 978-3-642-03414-5. doi:10.1007/978-3-642-03415-2

- Gillis, Justin (19 Sep 2012). "Ending Its Summer Melt, Arctic Sea Ice Sets a New Low That Leads to Warnings". The New York Times. Retrieved 5 Oct 2012

- Allen, John Eliot; Burns, Majorie; Sargent, Sam C. (1986). Cataclysms on the Columbia. Portland: Timber Press. ISBN 0-88192-215-3

- Rothrock, D.A.; Zhang, J. (2005). "Arctic Ocean Sea Ice Volume: What Explains Its Recent Depletion?". J. Geophys. Res. 110 (C1): C01002. Bibcode:2005JGRC..11001002R. doi:10.1029/2004JC002282

- Barry, Roger G.; Blanken, Peter D. (2016-05-23). Microclimate and Local Climate. Cambridge University Press. ISBN 9781316652336

- Coopes, Amy (22 December 2011). "Huge iceberg foils Mawson centenary plans". Cosmos. Archived from the original on February 28, 2014. Retrieved 23 February 2014

- Dorbolo, S,; Adami, N.; Dubois, C.; Caps, H.; Vandewalle, N.; Barbois-Texier, B. "Rotation of melting ice disks due to melt fluid flow". Phys. Rev. E. 93: 1–5. doi:10.1103/PhysRevE.93.03311

An Overview of Ice Age

An ice age is the period of extreme low temperatures on Earth. The three main types of evidence for ice age are geological, paleontological and chemical. The major ice age periods are Andean-Saharan, Cryogenian, Huronian, Quarternary glaciation and Karoo Ice Age. The aspects elucidated in this chapter are of vital importance, and provide a better understanding of ice age.

Ice Age

An artist's impression of ice age Earth at glacial maximum. Based on: *Crowley, T. J. (1995). "Ice age terrestrial carbon changes revisited".*

An ice age is a period of long-term reduction in the temperature of Earth's surface and atmosphere, resulting in the presence or expansion of continental and polar ice sheets and alpine glaciers. Within a long-term ice age, individual pulses of cold climate are termed "glacial periods" (or alternatively "glacials" or "glaciations" or colloquially as "ice age"), and intermittent warm periods are called "interglacials". In the terminology of glaciology, *ice age* implies the presence of extensive ice sheets in both northern and southern hemispheres. By this definition, we are in an interglacial period—the Holocene—of the ice age. The ice age began 2.6 million years ago at the start of the Pleistocene epoch, because the Greenland, Arctic, and Antarctic ice sheets still exist.

Origin of Ice Age Theory

In 1742 Pierre Martel (1706–1767), an engineer and geographer living in Geneva, visit-ed the valley of Chamonix in the Alps of Savoy. Two years later he published an account of his journey. He reported that the inhabitants of that valley attributed the dispersal of erratic boulders to the glaciers, saying that they had once extended much farther. Later similar explanations were reported from other regions of the Alps. In 1815 the carpen-ter and chamois hunter Jean-Pierre Perraudin (1767–1858) explained erratic boulders in the Val de Bagnes in the Swiss canton of Valais as being due to glaciers previously extending further. An unknown woodcutter from Meiringen in the Bernese Oberland advocated a similar idea in a discussion with the Swiss-German geologist Jean de Char-pentier (1786–1855) in 1834. Comparable explanations are also known from the Val de Ferret in the Valais and the Seeland in western Switzerland and in Goethe's scientific work. Such explanations could also be found in other parts of the world. When the Ba-varian naturalist Ernst von Bibra (1806–1878) visited the Chilean Andes in 1849–1850, the natives attributed fossil moraines to the former action of glaciers.

Meanwhile, European scholars had begun to wonder what had caused the dispersal of erratic material. From the middle of the 18th century, some discussed ice as a means of transport. The Swedish mining expert Daniel Tilas (1712–1772) was, in 1742, the first person to suggest drifting sea ice in order to explain the presence of erratic boulders in the Scandinavian and Baltic regions. In 1795, the Scottish philosopher and gentleman naturalist, James Hutton (1726–1797), explained erratic boulders in the Alps by the action of glaciers. Two decades later, in 1818, the Swedish botanist Göran Wahlenberg (1780–1851) published his theory of a glaciation of the Scandinavian peninsula. He regarded glaciation as a regional phenomenon.

The Antarctic ice sheet. Ice sheets expand during an ice age.

Only a few years later, the Danish-Norwegian geologist Jens Esmark (1762–1839) ar-gued a sequence of worldwide ice ages. In a paper published in 1824, Esmark proposed changes in climate as the cause of those glaciations. He attempted to show that they originated from changes in Earth's orbit. During the following years, Esmark's ideas were discussed and taken over in parts by Swedish, Scottish and German scientists. At the University of Edinburgh Robert Jameson (1774–1854) seemed to be relatively open to Esmark's ideas, as reviewed by Norwegian professor of glaciology Bjørn G. Andersen (1992). Jameson's remarks about ancient glaciers in Scotland were most probably prompted by Esmark. In Germany, Albrecht Reinhard Bernhardi (1797–1849), a geologist and professor of forestry at an academy in Dreissigacker, since incorporated in the southern Thuringian city of Meiningen, adopted Esmark's theory. In a paper published in 1832, Bernhardi speculated about former polar ice caps reaching as far as the temperate zones of the globe.

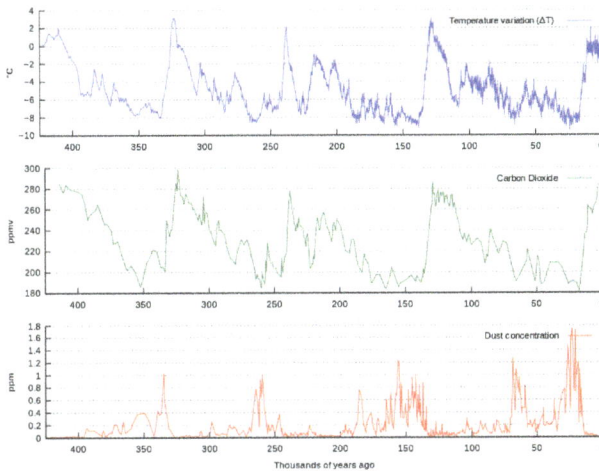

Variations in temperature, CO_2, and dust from the Vostok ice core over the last 400,000 years.

In 1829, independently of these debates, the Swiss civil engineer Ignaz Venetz (1788–1859) explained the dispersal of erratic boulders in the Alps, the nearby Jura Mountains, and the North German Plain as being due to huge glaciers. When he read his paper before the Schweizerische Naturforschende Gesellschaft, most scientists remained sceptical. Finally, Venetz convinced his friend Jean de Charpentier. De Charpentier transformed Venetz's idea into a theory with a glaciation limited to the Alps. His thoughts resembled Wahlenberg's theory. In fact, both men shared the same volcanistic, or in de Charpentier's case rather plutonistic assumptions, about the Earth's history. In 1834, de Charpentier presented his paper before the Schweizerische Naturforschende Gesellschaft. In the meantime, the German botanist Karl Friedrich Schimper (1803–1867) was studying mosses which were growing on erratic boulders in the alpine upland of Bavaria. He began to wonder where such masses of stone had come from. During the summer of 1835 he made some excursions to the Bavarian Alps. Schimper came to the conclusion that ice must have been the means of transport for the boulders in the alpine upland. In the winter of 1835 to 1836 he held some lectures in Munich. Schimper then assumed that there must have been global times of obliteration ("Verödungszeiten") with a cold climate and frozen water. Schimper spent the summer months of 1836 at Devens, near Bex, in the Swiss Alps with his former university friend Louis Agassiz (1801–1873) and Jean de Charpentier. Schimper, de Charpentier and possibly Venetz convinced Agassiz that there had been a time of glaciation. During the winter of 1836/37, Agassiz and Schimper developed the theory of a sequence of glaciations. They mainly drew upon the preceding works of Venetz, de Charpentier and on their own fieldwork. Agassiz appears to have been already familiar with Bernhardi's paper at that time. At the beginning of 1837, Schimper coined the term "ice age" (*"Eiszeit"*) for the period of the glaciers. In July 1837 Agassiz presented their synthesis before the annual meeting of the Schweizerische Naturforschende Gesellschaft at Neuchâtel. The audience was very critical and some opposed to the new theory because it contradicted the established opinions on climatic history. Most contemporary scientists thought that the Earth had been gradually cooling down since its birth as a molten globe.

In order to overcome this rejection, Agassiz embarked on geological fieldwork. He published his book *Study on Glaciers* ("Études sur les glaciers") in 1840. De Charpentier was put out by this, as he had also been preparing a book about the glaciation of the Alps. De Charpentier felt that Agassiz should have given him precedence as it was he who had introduced Agassiz to in-depth glacial research. Besides that, Agassiz had, as a result of personal quarrels, omitted any mention of Schimper in his book.

All together, it took several decades until the ice age theory was fully accepted by scientists. This happened on an international scale in the second half of the 1870s following the work of James Croll, including the publication of *Climate and Time, in Their Geological Relations* in 1875, which provided a credible explanation for the causes of ice ages.

Evidence for Ice Ages

There are three main types of evidence for ice ages: geological, chemical, and paleontological.

Geological evidence for ice ages comes in various forms, including rock scouring and scratching, glacial moraines, drumlins, valley cutting, and the deposition of till or tillites and glacial erratics. Successive glaciations tend to distort and erase the geological evidence, making it difficult to interpret. Furthermore, this evidence was difficult to date exactly; early theories assumed that the glacials were short compared to the long interglacials. The advent of sediment and ice cores revealed the true situation: glacials are long, interglacials short. It took some time for the current theory to be worked out.

The chemical evidence mainly consists of variations in the ratios of isotopes in fossils present in sediments and sedimentary rocks and ocean sediment cores. For the most recent glacial periods ice cores provide climate proxies from their ice, and atmospheric samples from included bubbles of air. Because water containing heavier isotopes has a higher heat of evaporation, its proportion decreases with colder conditions. This allows a temperature record to be constructed. However, this evidence can be confounded by other factors recorded by isotope ratios.

The paleontological evidence consists of changes in the geographical distribution of fossils. During a glacial period cold-adapted organisms spread into lower latitudes, and organisms that prefer warmer conditions become extinct or are squeezed into lower latitudes. This evidence is also difficult to interpret because it requires (1) sequences of sediments covering a long period of time, over a wide range of latitudes and which are easily correlated; (2) ancient organisms which survive for several million years without change and whose temperature preferences are easily diagnosed; and (3) the finding of the relevant fossils.

Despite the difficulties, analysis of ice core and ocean sediment cores has shown periods of glacials and interglacials over the past few million years. These also confirm the linkage between ice ages and continental crust phenomena such as glacial moraines, drumlins, and glacial erratics. Hence the continental crust phenomena are accepted as good evidence of earlier ice ages when they are found in layers created much earlier than the time range for which ice cores and ocean sediment cores are available.

Major Ice Ages

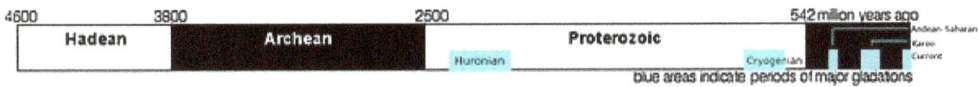

Timeline of glaciations, shown in blue.

There have been at least five major ice ages in the earth's past (the Huronian, Cryogenian, Andean-Saharan, Karoo Ice Age and the Quaternary glaciation). Outside these ages, the Earth seems to have been ice-free even in high latitudes.

Ice age map of northern Germany and its northern neighbours. Red: maximum limit of Weichselian glacial; yellow: Saale glacial at maximum (Drenthe stage); blue: Elster glacial maximum glaciation.

Rocks from the earliest well established ice age, called the Huronian, formed around 2.4 to 2.1 Ga (billion years) ago during the early Proterozoic Eon. Several hundreds of km of the Huronian Supergroup are exposed 10–100 km north of the north shore of Lake Huron extending from near Sault Ste. Marie to Sudbury, northeast of Lake

Huron, with giant layers of now-lithified till beds, dropstones, varves, outwash, and scoured basement rocks. Correlative Huronian deposits have been found near Marquette, Michigan, and correlation has been made with Paleoproterozoic glacial deposits from Western Australia.

The next well-documented ice age, and probably the most severe of the last billion years, occurred from 850 to 630 million years ago (the Cryogenian period) and may have produced a Snowball Earth in which glacial ice sheets reached the equator, possibly being ended by the accumulation of greenhouse gases such as CO_2 produced by volcanoes. "The presence of ice on the continents and pack ice on the oceans would inhibit both silicate weathering and photosynthesis, which are the two major sinks for CO_2 at present." It has been suggested that the end of this ice age was responsible for the subsequent Ediacaran and Cambrian explosion, though this model is recent and controversial.

The Andean-Saharan occurred from 460 to 420 million years ago, during the Late Ordovician and the Silurian period.

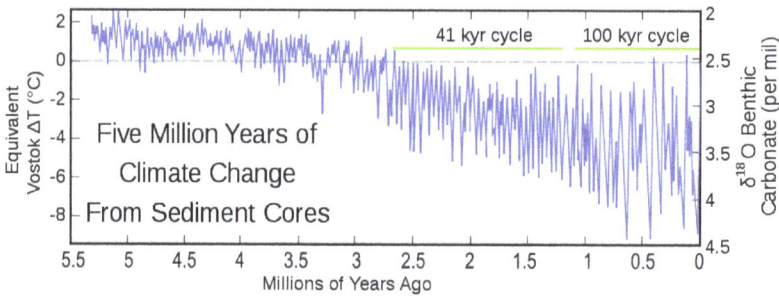

Sediment records showing the fluctuating sequences of glacials and interglacials during the last several million years.

The evolution of land plants at the onset of the Devonian period caused a long term increase in planetary oxygen levels and reduction of CO_2 levels, which resulted in the Karoo Ice Age. It is named after the glacial tills found in the Karoo region of South Africa, where evidence for this ice age was first clearly identified. There were extensive polar ice caps at intervals from 360 to 260 million years ago in South Africa during the Carboniferous and early Permian Periods. Correlatives are known from Argentina, also in the center of the ancient supercontinent Gondwanaland.

The current ice age, the Pliocene-Quaternary glaciation, started about 2.58 million years ago during the late Pliocene, when the spread of ice sheets in the Northern Hemisphere began. Since then, the world has seen cycles of glaciation with ice sheets advancing and retreating on 40,000- and 100,000-year time scales called glacial periods, glacials or glacial advances, and interglacial periods, interglacials or glacial retreats. The earth is currently in an interglacial, and the last glacial period ended about 10,000 years ago. All that remains of the continental ice sheets are the Greenland and Antarctic ice sheets and smaller glaciers such as on Baffin Island.

Ice ages can be further divided by location and time; for example, the names *Riss* (180,000–130,000 years bp) and *Würm* (70,000–10,000 years bp) refer specifically to glaciation in the Alpine region. The maximum extent of the ice is not maintained for the full interval. The scouring action of each glaciation tends to remove most of the evidence of prior ice sheets almost completely, except in regions where the later sheet does not achieve full coverage.

Evidence of a recent, extreme ice age on Mars was published by the journal *Science* in 2016. Just 370,000 years ago, the planet would have appeared more white than red.

Glacials and Interglacials

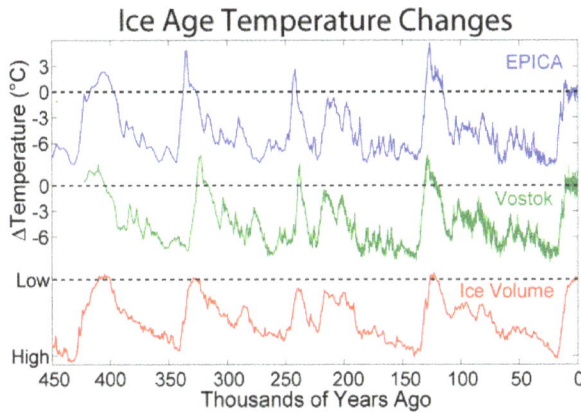

Shows the pattern of temperature and ice volume changes associated with recent glacials and interglacials.

Minimum and Maximum Glaciation

Within the ice ages (or at least within the current one), more temperate and more severe periods occur. The colder periods are called *glacial periods*, the warmer periods *interglacials*, such as the Eemian Stage.

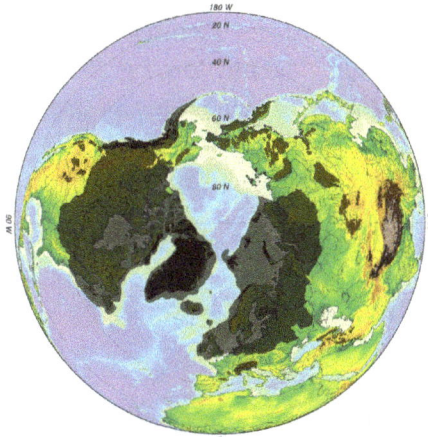

Minimum (interglacial, black) and maximum (glacial, grey) glaciation of the northern hemisphere.

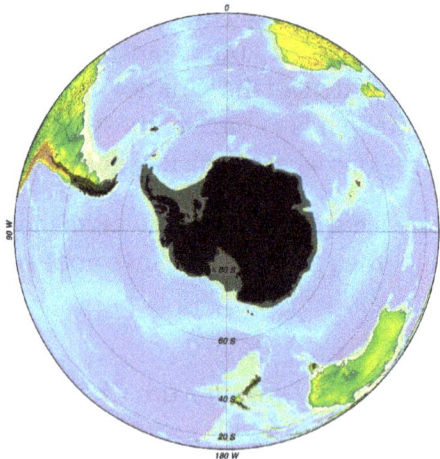

Minimum (interglacial, black) and maximum (glacial, grey)
glaciation of the southern hemisphere.

Glacials are characterized by cooler and drier climates over most of the earth and large land and sea ice masses extending outward from the poles. Mountain glaciers in otherwise unglaciated areas extend to lower elevations due to a lower snow line. Sea levels drop due to the removal of large volumes of water above sea level in the icecaps. There is evidence that ocean circulation patterns are disrupted by glaciations. Since the earth has significant continental glaciation in the Arctic and Antarctic, we are currently in a glacial minimum of a glaciation. Such a period between glacial maxima is known as an *interglacial*. The glacials and interglacials also coincided with changes in Earth's orbit called Milankovitch cycles.

The earth has been in an interglacial period known as the Holocene for more than 11,000 years. It was conventional wisdom that the typical interglacial period lasts about 12,000 years, but this has been called into question recently. For example, an article in *Nature* argues that the current interglacial might be most analogous to a previous interglacial that lasted 28,000 years. Predicted changes in orbital forcing suggest that the next glacial period would begin at least 50,000 years from now, even in absence of human-made global warming. Moreover, anthropogenic forcing from increased greenhouse gases might outweigh orbital forcing for as long as intensive use of fossil fuels continues.

Positive and Negative Feedback in Glacial Periods

Each glacial period is subject to positive feedback which makes it more severe, and negative feedback which mitigates and (in all cases so far) eventually ends it.

Positive Feedback Processes

Ice and snow increase Earth's albedo, i.e. they make it reflect more of the sun's energy and absorb less. Hence, when the air temperature decreases, ice and snow fields grow, and this continues until competition with a negative feedback mechanism forces the system to an equilibrium. Also, the reduction in forests caused by the ice's expansion increases albedo.

Another theory proposed by Ewing and Donn in 1956 hypothesized that an ice-free Arctic Ocean leads to increased snowfall at high latitudes. When low-temperature ice covers the Arctic Ocean there is little evaporation or sublimation and the polar regions are quite dry in terms of precipitation, comparable to the amount found in mid-latitude deserts. This low precipitation allows high-latitude snowfalls to melt during the summer. An ice-free Arctic Ocean absorbs solar radiation during the long summer days, and evaporates more water into the Arctic atmosphere. With higher precipitation, portions of this snow may not melt during the summer and so glacial ice can form at lower altitudes *and* more southerly latitudes, reducing the temperatures over land by increased albedo as noted above. Furthermore, under this hypothesis the lack of oceanic pack ice allows increased exchange of waters between the Arctic and the North Atlantic Oceans, warming the Arctic and cooling the North Atlantic. (Current projected consequences of global warming include a largely ice-free Arctic Ocean within 5–20 years) Additional fresh water flowing into the North Atlantic during a warming cycle may also reduce the global ocean water circulation. Such a reduction (by reducing the effects of the Gulf Stream) would have a cooling effect on northern Europe, which in turn would lead to increased low-latitude snow retention during the summer. It has also been suggested that during an extensive glacial, glaciers may move through the Gulf of Saint Lawrence, extending into the North Atlantic Ocean far enough to block the Gulf Stream.

Negative Feedback Processes

Ice sheets that form during glaciations cause erosion of the land beneath them. After some time, this will reduce land above sea level and thus diminish the amount of space on which ice sheets can form. This mitigates the albedo feedback, as does the lowering in sea level that accompanies the formation of ice sheets.

Another factor is the increased aridity occurring with glacial maxima, which reduces the precipitation available to maintain glaciation. The glacial retreat induced by this or any other process can be amplified by similar inverse positive feedbacks as for glacial advances.

According to research published in *Nature Geoscience*, human emissions of carbon dioxide (CO_2) will defer the next ice age. Researchers used data on Earth's orbit to find the historical warm interglacial period that looks most like the current one and from this have predicted that the next ice age would usually begin within 1,500 years. They go on to say that emissions have been so high that it will not.

Causes of Ice Ages

The causes of ice ages are not fully understood for either the large-scale ice age periods or the smaller ebb and flow of glacial–interglacial periods within an ice age. The consensus is that several factors are important: atmospheric composition, such as the concentrations of carbon dioxide and methane (the specific levels of the previously mentioned gases are now able to be seen with the new ice core samples from EPICA Dome C in Antarctica over the past 800,000 years) changes in the earth's orbit around the Sun known as Milankovitch cycles, the motion of tectonic plates resulting in changes in the relative location and amount of continental and oceanic crust on the earth's surface, which affect wind and ocean currents, variations in solar output, the orbital dynamics of the Earth–Moon system, the impact of relatively large meteorites and volcanism including eruptions of supervolcanoes.

Some of these factors influence each other. For example, changes in Earth's atmospheric composition (especially the concentrations of greenhouse gases) may alter the climate, while climate change itself can change the atmospheric composition (for example by changing the rate at which weathering removes CO_2).

Maureen Raymo, William Ruddiman and others propose that the Tibetan and Colorado Plateaus are immense CO_2 "scrubbers" with a capacity to remove enough CO_2 from the global atmosphere to be a significant causal factor of the 40 million year Cenozoic Cooling trend. They further claim that approximately half of their uplift (and CO_2 "scrubbing" capacity) occurred in the past 10 million years.

Changes in Earth's Atmosphere

There is evidence that greenhouse gas levels fell at the start of ice ages and rose during the retreat of the ice sheets, but it is difficult to establish cause and effect. Greenhouse gas levels may also have been affected by other factors which have been proposed as causes of ice ages, such as the movement of continents and volcanism.

The Snowball Earth hypothesis maintains that the severe freezing in the late Proterozoic was ended by an increase in CO_2 levels in the atmosphere, mainly from volcanoes, and some supporters of Snowball Earth argue that it was caused in the first place by a reduction in atmospheric CO_2. The hypothesis also warns of future Snowball Earths.

In 2009, further evidence was provided that changes in solar insolation provide the initial trigger for the earth to warm after an Ice Age, with secondary factors like increases in greenhouse gases accounting for the magnitude of the change.

Human-induced Changes

There is considerable evidence that over the very recent period of the last 100–1000 years, the sharp increases in human activity, especially the burning of fossil fuels, has caused the parallel sharp and accelerating increase in atmospheric greenhouse gases which trap the sun's heat. The consensus theory of the scientific community is that the resulting greenhouse effect is a principal cause of the increase in global warming which has occurred over the same period, and a chief contributor to the accelerated melting of the remaining glaciers and polar ice. A 2012 investigation finds that dinosaurs released methane through digestion in a similar amount to humanity's current methane release, which "could have been a key factor" to the very warm climate 150 million years ago.

William Ruddiman has proposed the early anthropocene hypothesis, according to which the anthropocene era, as some people call the most recent period in the earth's history when the activities of the human species first began to have a significant global impact on the earth's climate and ecosystems, did not begin in the 18th century with the advent of the Industrial Era, but dates back to 8,000 years ago, due to intense farming activities of our early agrarian ancestors. It was at that time that atmospheric greenhouse gas concentrations stopped following the periodic pattern of the Milankovitch cycles. In his overdue-glaciation hypothesis Ruddiman states that an incipient glacial would probably have begun several thousand years ago, but the arrival of that scheduled glacial was forestalled by the activities of early farmers.

At a meeting of the American Geophysical Union (December 17, 2008), scientists detailed evidence in support of the controversial idea that the introduction of large-scale rice agriculture in Asia, coupled with extensive deforestation in Europe began to alter world climate by pumping significant amounts of greenhouse gases into the atmosphere over the last 1,000 years. In turn, a warmer atmosphere heated the oceans making them much less efficient storehouses of carbon dioxide and reinforcing global warming, possibly forestalling the onset of a new glacial age.

Position of the Continents

The geological record appears to show that ice ages start when the continents are in positions which block or reduce the flow of warm water from the equator to the poles and thus allow ice sheets to form. The ice sheets increase Earth's reflectivity and thus reduce the absorption of solar radiation. With less radiation absorbed the atmosphere cools; the cooling allows the ice sheets to grow, which further increases reflectivity in a positive feedback loop. The ice age continues until the reduction in weathering causes an increase in the greenhouse effect.

There are three known configurations of the continents which block or reduce the flow of warm water from the equator to the poles:

- A continent sits on top of a pole, as Antarctica does today.

- A polar sea is almost land-locked, as the Arctic Ocean is today.

- A supercontinent covers most of the equator, as Rodinia did during the Cryogenian period.

Since today's Earth has a continent over the South Pole and an almost land-locked ocean over the North Pole, geologists believe that Earth will continue to experience glacial periods in the geologically near future.

Some scientists believe that the Himalayas are a major factor in the current ice age, because these mountains have increased Earth's total rainfall and therefore the rate at which carbon dioxide is washed out of the atmosphere, decreasing the greenhouse effect. The Himalayas' formation started about 70 million years ago when the Indo-Australian Plate collided with the Eurasian Plate, and the Himalayas are still rising by about 5 mm per year because the Indo-Australian plate is still moving at 67 mm/year. The history of the Himalayas broadly fits the long-term decrease in Earth's average temperature since the mid-Eocene, 40 million years ago.

Fluctuations in Ocean Currents

Another important contribution to ancient climate regimes is the variation of ocean currents, which are modified by continent position, sea levels and salinity, as well as other factors. They have the ability to cool (e.g. aiding the creation of Antarctic ice) and the ability to warm (e.g. giving the British Isles a temperate as opposed to a boreal climate). The closing of the Isthmus of Panama about 3 million years ago may have ushered in the present period of strong glaciation over North America by ending the exchange of water between the tropical Atlantic and Pacific Oceans.

Analyses suggest that ocean current fluctuations can adequately account for recent glacial oscillations. During the last glacial period the sea-level has fluctuated 20–30 m as water was sequestered, primarily in the Northern Hemisphere ice sheets. When ice collected and the sea level dropped sufficiently, flow through the Bering Strait (the narrow strait between Siberia and Alaska is about 50 m deep today) was reduced, resulting in increased flow from the North Atlantic. This realigned the thermohaline circulation in the Atlantic, increasing heat transport into the Arctic, which melted the polar ice accumulation and reduced other continental ice sheets. The release of water raised sea levels again, restoring the ingress of colder water from the Pacific with an accompanying shift to northern hemisphere ice accumulation.

Uplift of the Tibetan Plateau

Matthias Kuhle's geological theory of Ice Age development was suggested by the existence of an ice sheet covering the Tibetan Plateau during the Ice Ages (Last Glacial Maximum?). According to Kuhle, the plate-tectonic uplift of Tibet past the snow-line has led to a surface of c. 2,400,000 square kilometres (930,000 sq mi) changing from bare land to ice with a 70% greater albedo. The reflection of energy into space resulted in a global cooling, triggering the Pleistocene Ice Age. Because this highland is at a subtropical latitude, with 4 to 5 times the insolation of high-latitude areas, what would be Earth's strongest heating surface has turned into a cooling surface.

Kuhle explains the interglacial periods by the 100,000-year cycle of radiation changes due to variations in Earth's orbit. This comparatively insignificant warming, when combined with the lowering of the Nordic inland ice areas and Tibet due to the weight of the superimposed ice-load, has led to the repeated complete thawing of the inland ice areas.

Variations in Earth's Orbit (Milankovitch Cycles)

The Milankovitch cycles are a set of cyclic variations in characteristics of the Earth's orbit around the Sun. Each cycle has a different length, so at some times their effects reinforce each other and at other times they (partially) cancel each other.

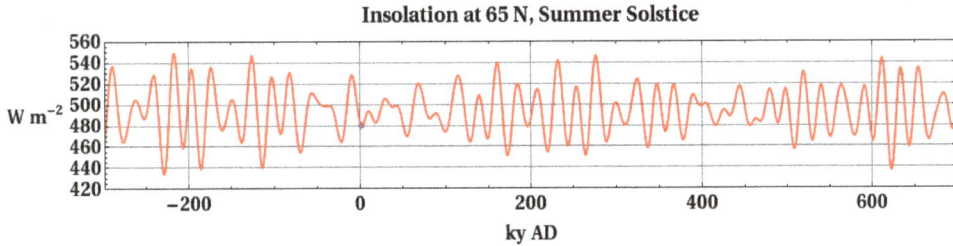

Insolation at 65 N, Summer Solstice

Past and future of daily average insolation at top of the atmosphere on the day of the summer solstice, at 65 N latitude.

There is strong evidence that the Milankovitch cycles affect the occurrence of glacial and interglacial periods within an ice age. The present ice age is the most studied and best understood, particularly the last 400,000 years, since this is the period covered by ice cores that record atmospheric composition and proxies for temperature and ice volume. Within this period, the match of glacial/interglacial frequencies to the Milanković orbital forcing periods is so close that orbital forcing is generally accepted. The combined effects of the changing distance to the Sun, the precession of the Earth's axis, and the changing tilt of the Earth's axis redistribute the sunlight received by the Earth. Of particular importance are changes in the tilt of the Earth's axis, which affect the intensity of seasons. For example, the amount of solar influx in July at 65 degrees north latitude varies by as much as 22% (from 450 W/m² to 550 W/m²). It is widely believed that ice sheets advance when summers become too cool to melt all of the accumulated snowfall from the previous winter. Some believe that the strength of the orbital forcing is too small to trigger glaciations, but feedback mechanisms like CO_2 may explain this mismatch.

While Milankovitch forcing predicts that cyclic changes in the Earth's orbital elements can be expressed in the glaciation record, additional explanations are necessary to explain which cycles are observed to be most important in the timing of glacial–interglacial periods. In particular, during the last 800,000 years, the dominant period of glacial–interglacial oscillation has been 100,000 years, which corresponds to changes in Earth's orbital eccentricity and orbital inclination. Yet this is by far the weakest of the three frequencies predicted by Milankovitch. During the period 3.0–0.8 million years ago, the dominant pattern of glaciation corresponded to the 41,000-year period of changes in Earth's obliquity (tilt of the axis). The reasons for dominance of one frequency versus another are poorly understood and an active area of current research, but the answer probably relates to some form of resonance in the Earth's climate system. Recent work suggests that the 100K year cycle dominates due to increased southern-pole sea-ice increasing total solar reflectivity.

The "traditional" Milankovitch explanation struggles to explain the dominance of the 100,000-year cycle over the last 8 cycles. Richard A. Muller, Gordon J. F. MacDonald, and others have pointed out that those calculations are for a two-dimensional orbit of Earth but the three-dimensional orbit also has a 100,000-year cycle of orbital inclination. They proposed that these variations in orbital inclination lead to variations in insolation, as the Earth moves in and out of known dust bands in the solar system. Although this is a different mechanism to the traditional view, the "predicted" periods over the last 400,000 years are nearly the same. The Muller and MacDonald theory, in turn, has been challenged by Jose Antonio Rial.

Another worker, William Ruddiman, has suggested a model that explains the 100,000-year cycle by the modulating effect of eccentricity (weak 100,000-year cycle) on precession (26,000-year cycle) combined with greenhouse gas feedbacks in the 41,000- and 26,000-year cycles. Yet another theory has been advanced by Peter Huybers who argued that the 41,000-year cycle has always been dominant, but that the Earth has entered a mode of climate behavior where only the second or third cycle triggers an ice age. This would imply that the 100,000-year periodicity is really an illusion created by averaging together cycles lasting 80,000 and 120,000 years. This theory is consistent with a simple empirical multi-state model proposed by Didier Paillard. Paillard suggests that the late Pleistocene glacial cycles can be seen as jumps between three quasi-stable climate states. The jumps are induced by the orbital forcing, while in the early Pleistocene the 41,000-year glacial cycles resulted from jumps between only two climate states. A dynamical model explaining this behavior was proposed by Peter Ditlevsen. This is in support of the suggestion that the late Pleistocene glacial cycles are not due to the weak 100,000-year eccentricity cycle, but a non-linear response to mainly the 41,000-year obliquity cycle.

Variations in the Sun's Energy Output

There are at least two types of variation in the Sun's energy output

- In the very long term, astrophysicists believe that the Sun's output increases by about 7% every one billion (10^9) years.

- Shorter-term variations such as sunspot cycles, and longer episodes such as the Maunder Minimum, which occurred during the coldest part of the Little Ice Age.

The long-term increase in the Sun's output cannot be a cause of ice ages.

Volcanism

Volcanic eruptions may have contributed to the inception and/or the end of ice age periods. At times during the paleoclimate, carbon dioxide levels were two or three times greater than today. Volcanoes and movements in continental plates contributed to high amounts of CO_2 in the atmosphere. Carbon dioxide from volcanoes probably contributed to periods with highest overall temperatures. One suggested explanation of the Paleocene-Eocene Thermal Maximum is that undersea volcanoes released methane from clathrates and thus caused a large and rapid increase in the greenhouse effect. There appears to be no geological evidence for such eruptions at the right time, but this does not prove they did not happen.

Recent Glacial and Interglacial Phases

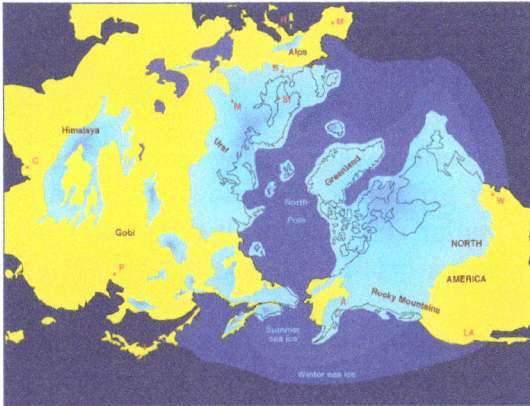

Northern hemisphere glaciation during the last ice ages. The setup of 3 to 4 kilometer thick ice sheets caused a sea level lowering of about 120 m.

Glacial Stages in North America

The major glacial stages of the current ice age in North America are the Illinoian, Eemian and Wisconsin glaciation. The use of the Nebraskan, Afton, Kansan, and Yarmouthian stages to subdivide the ice age in North America have been discontinued by Quaternary geologists and geomorphologists. These stages have all been merged into the Pre-Illinoian in the 1980s.

During the most recent North American glaciation, during the latter part of the Last Glacial Maximum (26,000 to 13,300 years ago), ice sheets extended to about 45th parallel north. These sheets were 3 to 4 kilometres (1.9 to 2.5 mi) thick.

This Wisconsin glaciation left widespread impacts on the North American landscape. The Great Lakes and the Finger Lakes were carved by ice deepening old valleys. Most of the lakes in Minnesota and Wisconsin were gouged out by glaciers and later filled with glacial meltwaters. The old Teays River drainage system was radically altered and largely reshaped into the Ohio River drainage system. Other rivers were dammed and diverted to new channels, such as Niagara Falls, which formed a dramatic waterfall and gorge, when the waterflow encountered a limestone escarpment. Another similar waterfall, at the present Clark Reservation State Park near Syracuse, New York, is now dry.

The area from Long Island to Nantucket, Massachusetts was formed from glacial till, and the plethora of lakes on the Canadian Shield in northern Canada can be almost entirely attributed to the action of the ice. As the ice retreated and the rock dust dried, winds carried the material hundreds of miles, forming beds of loess many dozens of feet thick in the Missouri Valley. Post-glacial rebound continues to reshape the Great Lakes and other areas formerly under the weight of the ice sheets.

The Driftless Area, a portion of western and southwestern Wisconsin along with parts of adjacent Minnesota, Iowa, and Illinois, was not covered by glaciers.

Last Glacial Period

A specially interesting climatic change during glacial times has taken place in the semi-arid Andes. Beside the expected cooling down in comparison with the current climate, a significant precipitation change happened here. So, researches in the presently semiarid subtropic Aconcagua-massif (6,962 m) have shown an unexpectedly extensive glacial glaciation of the type "ice stream network". The connected valley glaciers exceeding 100 km in length, flowed down on the East-side of this section of the Andes at 32–34°S and 69–71°W as far as a height of 2,060 m and on the western luff-side still clearly deeper. Where current glaciers scarcely reach 10 km in length, the snowline (ELA) runs at a height of 4,600 m and at that time was lowered to 3,200 m asl, i.e. about 1,400 m. From this follows that—beside of an annual depression of temperature about c. 8.4 °C— here was an increase in precipitation. Accordingly, at glacial times the humid climatic belt that today is situated several latitude degrees further to the S, was shifted much further to the N.

Effects of Glaciation

Scandinavia exhibits some of the typical effects of ice age glaciation such as fjords and lakes.

Although the last glacial period ended more than 8,000 years ago, its effects can still be felt today. For example, the moving ice carved out the landscape in Canada, Greenland, northern Eurasia and Antarctica. The erratic boulders, till, drumlins, eskers, fjords, kettle lakes, moraines, cirques, horns, etc., are typical features left behind by the glaciers.

The weight of the ice sheets was so great that they deformed the Earth's crust and mantle. After the ice sheets melted, the ice-covered land rebounded. Due to the high viscosity of the Earth's mantle, the flow of mantle rocks which controls the rebound process is very slow—at a rate of about 1 cm/year near the center of rebound area today.

During glaciation, water was taken from the oceans to form the ice at high latitudes, thus global sea level dropped by about 110 meters, exposing the continental shelves and forming land-bridges between land-masses for animals to migrate. During deglaciation, the melted ice-water returned to the oceans, causing sea level to rise. This process can cause sudden shifts in coastlines and hydration systems resulting in newly submerged lands, emerging lands, collapsed ice dams resulting in salination of lakes, new ice dams creating vast areas of freshwater, and a general alteration in regional weather patterns on a large but temporary scale. It can even cause temporary reglaciation. This type of chaotic pattern of rapidly changing land, ice, saltwater and freshwater has been proposed as the likely model for the Baltic and Scandinavian regions, as well as much of central North America at the end of the last glacial maximum, with the present-day coastlines only being achieved in the last few millennia of prehistory. Also, the effect of elevation on Scandinavia submerged a vast continental plain that had existed under much of what is now the North Sea, connecting the British Isles to Continental Europe.

The redistribution of ice-water on the surface of the Earth and the flow of mantle rocks causes changes in the gravitational field as well as changes to the distribution of the moment of inertia of the Earth. These changes to the moment of inertia result in a change in the angular velocity, axis, and wobble of the Earth's rotation.

The weight of the redistributed surface mass loaded the lithosphere, caused it to flex and also induced stress within the Earth. The presence of the glaciers generally suppressed the movement of faults below. However, during deglaciation, the faults experience accelerated slip triggering earthquakes. Earthquakes triggered near the ice margin may in turn accelerate ice calving and may account for the Heinrich events. As more ice is removed near the ice margin, more intraplate earthquakes are induced and this positive feedback may explain the fast collapse of ice sheets.

In Europe, glacial erosion and isostatic sinking from weight of ice made the Baltic Sea, which before the Ice Age was all land drained by the Eridanos River.

Glacial Period

A glacial period (alternatively glacial or glaciation) is an interval of time (thousands of years) within an ice age that is marked by colder temperatures and glacier advances. Interglacials, on the other hand, are periods of warmer climate between glacial periods. The last glacial period ended about 15,000 years ago. The Holocene epoch is the current interglacial. A time when there are no glaciers on Earth is considered a greenhouse climate state.

Quaternary Ice Age

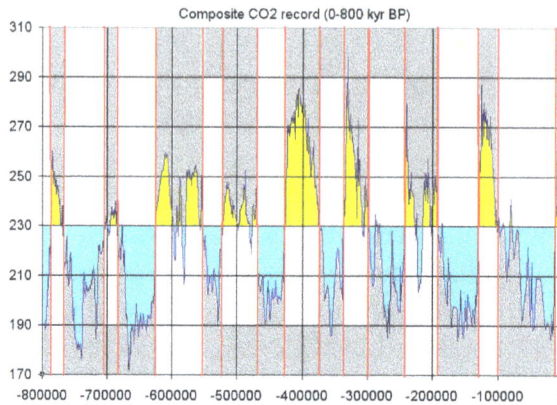

Glacial and interglacial cycles as represented by atmospheric CO_2, measured from ice core samples going back 800,000 years. The stage names are part of the North American and the European Alpine subdivisions. The correlation between both subdivisions is tentative.

Within the Quaternary glaciation (1.8 Ma to present), there have been a number of glacials and interglacials.

Last Glacial Period

The last glacial period was the most recent glacial period within the current ice age, occurring in the Pleistocene epoch, which began about 110,000 years ago and ended about 15,000 years ago. The glaciations that occurred during this glacial period covered many areas of the Northern Hemisphere and have different names, depending on their geographic distributions: *Wisconsin* (in North America), *Devensian* (in Great Britain), *Midlandian* (in Ireland), *Würm* (in the Alps), *Weichsel* (in northern central Europe), *Dali* (in East China), *Beiye* (in North China), *Taibai* (in Shaanxi) *Luojishan* (in Southwest Sichuan), *Zagunao* (in Northwest Sichuan), *Tianchi* (in Tianshan Mountains) *Qomolangma* (in Himalaya), and *Llanquihue* (in Chile). The glacial advance reached its maximum extent about 18,000 BP. In Europe, the ice sheet reached northern Germany.

Next Glacial Period

Since orbital variations are predictable, computer models that relate orbital variations to climate can predict future climate possibilities. Two caveats are necessary: that anthropogenic effects (human-assisted global warming) are likely to exert a larger influence over the short term; and that the mechanism by which orbital forcing influences climate is not well understood. Work by Berger and Loutre suggests that the current warm climate may last another 50,000 years.

Huronian Glaciation

The Huronian glaciation (or Makganyene glaciation) was a glaciation that extended from 2.4 billion years ago (Ga) to 2.1 Ga, during the Siderian and Rhyacian periods of the Paleoproterozoic era. The Huronian glaciation followed after the Great Oxygenation Event (GOE), a time when increased atmospheric oxygen decreased atmospheric methane. The oxygen combined with the methane to form carbon dioxide and water, which does not retain heat as well as methane does.

It is the oldest and longest ice age, occurring at a time when, in a biological sense, only simple, unicellular life existed on Earth. This ice age led to a mass-extinction on Earth.

Name Origin

This geological era was named for two non-glacial sediment deposits found between three separate horizons of glacial deposits of the Huronian Supergroup deposited between 2.5 and 2.2 billion years ago in the geographic area of Lake Huron.

Geological Context

The tectonic setting was one of a rifting continental margin. New continental crust would have resulted in chemical weathering. That coupled with reduced solar luminosity would have caused an 'antigreenhouse' effect as carbon dioxide in Earth's atmosphere was reduced. Volcanic sources in turn would have replenished that carbon dioxide, resulting in warming and interglacial periods. The Gowganda formation (2.3 Ga) contains "the most widespread and most convincing glaciogenic deposits of this era," according to Eyles and Young. Similar deposits are found in Michigan (2.1-2 Ga), the Black Hills (2.6-1.6 Ga), Chibougamau, Canadian Northern Territories (2.1 Ga) and Wyoming. Similar age deposits occur in the Griquatown Basin (2.3 Ga), India (1.8 Ga) and Australia (2.5-2.0 Ga).

Causes

Before the Huronian Ice Age, most organisms were anaerobic, but around this time, the cyanobacteria evolved photosynthesis. These bacteria were able to reproduce at exponential rates due to their new ecological niche, exploiting the near-limitless energy of the sun. Their photosynthesis produced oxygen as a waste product expelled into the air. At first, most of this oxygen was absorbed through the oxidization of surface iron and the decomposition of life forms. However, as the population of the cyanobacteria continued to grow, these oxygen-sinks became saturated. This led to a mass extinction of most life forms, which were anaerobic, as oxygen was toxic to them. As oxygen "polluted" the mostly methane atmosphere, and methane bonded with oxygen to form carbon dioxide and water, a different, thinner atmosphere emerged, and Earth began to lose heat. Thus began the Huronian Ice Age.

Andean-Saharan Glaciation

The Andean-Saharan glaciation occurred during the Paleozoic from 450 Ma to 420 Ma, during the late Ordovician and the Silurian period.

According to Eyles and Young, "A major glacial episode at c. 440 Ma, is recorded in Late Ordovician strata (predominantly Ashgillian) in West Africa (Tamadjert Formation of the Sahara), in Morocco (Tindouf Basin) and in west-central Saudi Arabia, all areas at polar latitudes at the time. From the Late Ordovician to the Early Silurian the centre of glaciation moved from northern Africa to southwestern South America."

During this period glaciation is known from Arabia, Sahara, West Africa, the south Amazon, and the Andes. The center of glaciation migrated from Sahara in the Ordovician (450-440 Ma) to South America in the Silurian (440-420 Ma). The maximum extent of glaciation developed in Africa and eastern Brazil.

A minor ice age, the Andean-Saharan was preceded by the Cryogenian ice ages (850-630 Ma, the Sturtian and Marinoan glaciations), often referred to as Snowball Earth, and followed by the Karoo Ice Age (350-260 Ma).

Karoo Ice Age

The Karoo Ice Age from 360–260 million years ago (Mya) was the second major ice age of the Phanerozoic Eon. It is named after the tillite (Dwyka Group) found in the Karoo region of South Africa (and adjacent areas), where evidence for this ice age was first clearly identified in the 19th century.

The tectonic assembly of the continents of Euramerica (later with the Uralian orogeny, into Laurasia) and Gondwana into Pangaea, in the Hercynian-Alleghany Orogeny, made a major continental land mass within the Antarctic region, and the closure of the Rheic Ocean and Iapetus Ocean saw disruption of warm-water currents in the Panthalassa Ocean and Paleotethys Sea, which led to progressive cooling of summers, and the snowfields accumulating in winters, causing mountainous alpine glaciers to grow, and then spread out of highland areas, making continental glaciers which spread to cover much of Gondwana.

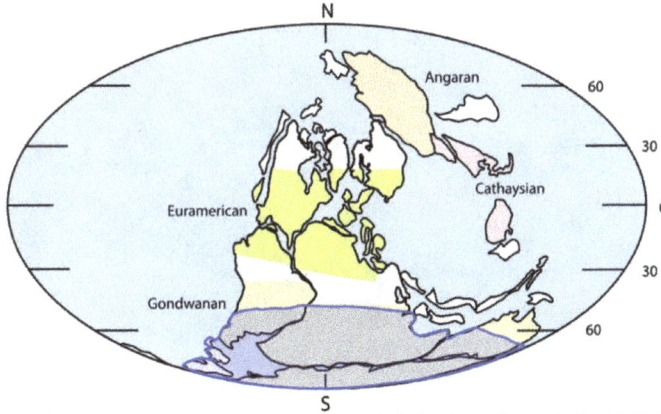

Approximate extent of the Karoo Glaciation (in blue),
over the Gondwana supercontinent during the Carboniferous and Permian periods

At least two major periods of glaciation have been discovered:

- The first glacial period was associated with the Mississippian subperiod (359.2–318.1 Mya): ice sheets expanded from a core in southern Africa and South America.

- The second glacial period was associated with the Pennsylvanian subperiod (318.1–299 Mya); ice sheets expanded from a core in Australia and India.

The extent of ancient glaciations in Antarctica is not well known, because the present ice sheet hides the evidence.

Late Paleozoic Glaciations

According to Eyles and Young, "Renewed Late Devonian glaciation is well documented in three large intracratonic basins in Brazil (Solimoes, Amazonas and Paranaiba basins) and in Bolivia. By the Early Carboniferous (c. 350 Ma) glacial strata were beginning to accumulate in sub-andean basins of Bolivia, Argentina and Paraguay. By the mid-Carboniferous glaciation had spread to Antarctica, Australia, southern Africa, the Indian Subcontinent, Asia and the Arabian Peninsula. During the Late Carboniferous glacial accumulation (c. 300 Ma) a very large area of Gondwana land mass was experiencing glacial conditions. The thickest glacial deposits of Permo-Carboniferous age are the Dwyka Formation (1000 m thick) in the Karoo Basin in southern Africa, the Itarare Group of the Parana Basin, Brazil (1400 m) and the Carnarvon Basin in eastern Australia. The Permo-Carboniferous glaciations are significant because of the marked glacio-eustatic changes in sea level that resulted and which are recorded in non-glacial basins. Late Paleozoic glaciation of Gondwana could be explained by the migration of the supercontinent across the South Pole."

Causes of the Karoo Ice Age

Glacial striations formed by Karoo Ice Age glaciers in the Witmarsum Colony, Paraná Basin, Paraná, Brazil.

The evolution of land plants with the onset of the Devonian Period, began a long-term increase in planetary oxygen levels. Large tree ferns, growing to 20 m high, were secondarily dominant to the large arborescent lycopods (30–40 m high) of the Carboniferous coal forests that flourished in equatorial swamps stretching from Appalachia to Poland, and later on the flanks of the Urals. Oxygen levels reached up to 35%, and global carbon dioxide got below the 300 parts per million level, which is today associated with glacial periods. This reduction in the greenhouse effect was coupled with lignin and cellulose (as tree trunks and other vegetation debris) accumulating and being buried in the great Carboniferous Coal Measures. The reduction of carbon dioxide levels in the atmosphere would be enough to begin the process of changing polar climates, leading to cooler summers which could not melt the previous winter's snow accumulations. The growth in snowfields to 6 m deep would create sufficient pressure to convert the lower levels to ice.

Earth's increased planetary albedo produced by the expanding ice sheets would lead to positive feedback loops, spreading the ice sheets still further, until the process hit limit. Falling global temperatures would eventually limit plant growth, and the rising levels of oxygen would increase the frequency of fire-storms because damp plant matter could burn. Both these effects return carbon dioxide to the atmosphere, reversing the "snowball" effect and forcing greenhouse warming, with CO_2 levels rising to 300 ppm in the following Permian period. Over a longer period the evolution of termites, whose stomachs provided an anoxic environment for methanogenic lignin-digesting bacteria, prevented further burial of carbon, returning carbon to the air as the greenhouse gas methane.

Once these factors brought a halt and a small reversal in the spread of ice sheets, the lower planetary albedo resulting from the fall in size of the glaciated areas would have been enough for warmer summers and winters and thus limit the depth of snowfields in areas from which the glaciers expanded. Rising sea levels produced by global warming drowned the large areas of flatland where previously anoxic swamps assisted in burial and removal of carbon (as coal). With a smaller area for deposition of carbon, more carbon dioxide was returned to the atmosphere, further warming the planet. By 250 Mya, planet Earth had returned to a percentage of oxygen similar to that found today.

The Effects of the Karoo Ice Age

The rising levels of oxygen in the Karoo Ice Age had major effects upon evolution of plants and animals. Higher oxygen concentration (and accompanying higher atmospheric pressure) enabled energetic metabolic processes which encouraged evolution of large land-dwelling vertebrates and flight, with the dragonfly-like *Meganeura*, an aerial predator, with a wingspan of 60 to 75 cm.

The harmless stocky-bodied and armoured millipede-like *Arthropleura* was 1.8 m long, and the semiterrestrial Hibbertopterid eurypterids were perhaps as large, and some scorpions reached 50 or 70 cm.

The rising levels of oxygen also led to the evolution of greater fire resistance in vegetation and ultimately to the evolution of flowering plants. Genetic studies have shown this was when angiosperms separated from cycads and gymnosperms.

In addition, the Karoo Ice Age has unique sedimentary sequences called cyclothems. These were produced by the repeated alterations of marine and nonmarine environments.

References

- Warren, John K. (2006). Evaporites: sediments, resources and hydrocarbons. Birkhäuser. p. 289. ISBN 978-3-540-26011-0

- Lockwood, J.G.; van Zinderen-Bakker, E. M. (November 1979). "The Antarctic Ice-Sheet: Regulator of Global Climates?: Review". The Geographical Journal. 145 (3): 469–471. JSTOR 633219. doi:10.2307/633219

- Tang, Haoshu; Chen, Yanjing (1 September 2013). "Global glaciations and atmospheric change at ca. 2.3 Ga". Geoscience Frontiers. 4 (5): 583–596. doi:10.1016/j.gsf.2013.02.003

- Eyles, Nicholas; Young, Grant (1994). Deynoux, M.; Miller, J.M.G.; Domack, E.W.; Eyles, N.; Fairchild, I.J.; Young, G.M., eds. Geodynamic controls on glaciation in Earth history, in Earth's Glacial Record. Cambridge: Cambridge University Press. pp. 10–18. ISBN 0521548039

- Kuhle, Matthias (December 1988). "Tibet and High-Asia: Results of the Sino-German Joint Expeditions (I)". GeoJournal. 17 (4): 581–595. JSTOR 41144345

- Hunt, A.G.; Malin, P.E.; Malin (14 May 1998). "Possible triggering of Heinrich events by ice-load-induced earthquakes". Nature. 393 (6681): 155–8. Bibcode:1998Natur.393..155H. doi:10.1038/30218

- Johnston, A. (1989). "The effect of large ice sheets on earthquake genesis". In Gregersen, S.; Basham, P. Earthquakes at North-Atlantic passive margins: Neotectonics and postglacial rebound. Dordrecht: Kluwer. pp. 581–599. ISBN 0-7923-0150-1

- Kuhle, Matthias (June 1987). "Subtropical Mountain- and Highland-Glaciation as Ice Age Triggers and the Waning of the Glacial Periods in the Pleistocene". GeoJournal. 14 (4): 393–421. JSTOR 41144132. doi:10.1007/BF02602717

- Andersen, Bjørn G. (1992). "Jens Esmark—a pioneer in glacial geology". Boreas. 21: 97–102. doi:10.1111/j.1502-3885.1992.tb00016.x

- Högele, M. A. (2011), Metastability of the Chafee-Infante equation with small heavy-tailed Lévy Noise (PDF), Humboldt-Universität zu Berlin, Mathematisch-Naturwissenschaftliche Fakultät II, retrieved 7 November 2015

- Ruddiman, William F. (2003). "The Anthropogenic Greenhouse Era Began Thousands of Years Ago" (PDF). Climatic Change. 61 (3): 261–293. doi:10.1023/B:CLIM.0000004577.17928.fa

Global Warming and its Effects on Glaciers

Global warming is the rise in temperatures caused by the release of carbon dioxide and greenhouse gasses into the atmosphere. The rise in temperature caused because of global change has resulted in rising sea levels, melting glaciers, severe droughts, etc. The topics discussed in the chapter are of great importance to broaden the existing knowledge on global warming.

Global Warming

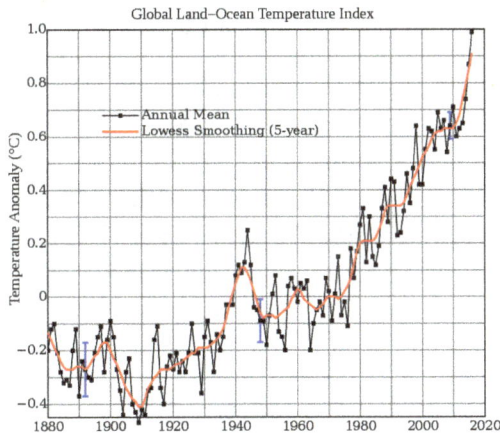

Global mean surface-temperature change from 1880 to 2016, relative to the 1951–1980 mean. The black line is the global annual mean and the red line is the five-year lowess smooth. The blue uncertainty bars show a 95% confidence interval.

Global warming, also referred to as climate change, is the observed century-scale rise in the average temperature of the Earth's climate system and its related effects. Multiple lines of scientific evidence show that the climate system is warming. Many of the observed changes since the 1950s are unprecedented in the instrumental temperature record which extends back to the mid 19th century, and in paleoclimate proxy records over thousands of years.

In 2013, the Intergovernmental Panel on Climate Change (IPCC) Fifth Assessment Report concluded that "It is *extremely likely* that human influence has been the dominant cause of the observed warming since the mid-20th century." The largest human influence has been emission of greenhouse gases such as carbon dioxide, methane and nitrous oxide. Climate model projections summarized in the report indicated that during the 21st century the global surface temperature is likely to rise a further 0.3 to 1.7 °C (0.5 to 3.1 °F) for their lowest emissions scenario and 2.6 to 4.8 °C (4.7 to 8.6 °F) for the highest emissions scenario. These findings have been recognized by the national science academies of the major industrialized nations and are not disputed by any scientific body of national or international standing.

Future climate change and associated impacts will differ from region to region around the globe. Anticipated effects include warming global temperature, rising sea levels, changing precipitation, and expansion of deserts in the subtropics. Warming is expected to be greater over land than over the oceans and greatest in the Arctic, with the continuing retreat of glaciers, permafrost and sea ice. Other likely changes include more frequent extreme weather events including heat waves, droughts, heavy rainfall with floods and heavy snowfall; ocean acidification; and species extinctions due to shifting temperature regimes. Effects significant to humans include the threat to food security from decreasing crop yields and the abandonment of populated areas due to rising sea levels. Because the climate system has a large "inertia" and greenhouse gases will stay in the atmosphere for a long time, many of these effects will not only exist for decades or centuries, but will persist for tens of thousands of years to come.

Possible societal responses to global warming include mitigation by emissions reduction, adaptation to its effects, building systems resilient to its effects, and possible future climate engineering. Most countries are parties to the United Nations Framework Convention on Climate Change (UNFCCC), whose ultimate objective is to prevent dangerous anthropogenic climate change. Parties to the UNFCCC have agreed that deep cuts in emissions are required and that global warming should be limited to well below 2.0 °C (3.6 °F) relative to pre-industrial levels, with efforts made to limit warming to 1.5 °C (2.7 °F).

Public reactions to global warming and concern about its effects are also increasing. A global 2015 Pew Research Center report showed a median of 54% consider it "a very serious problem". There are significant regional differences, with Americans and Chinese (whose economies are responsible for the greatest annual CO_2 emissions) among the least concerned.

Observed Temperature Changes

The global average (land and ocean) surface temperature shows a warming of 0.85 [0.65 to 1.06] °C in the period 1880 to 2012, based on multiple independently produced datasets. Earth's average surface temperature rose by 0.74±0.18 °C over the period 1906–2005. The rate of warming almost doubled for the last half of that period (0.13±0.03 °C per decade, versus 0.07±0.02 °C per decade). Although the increase of near-surface atmospheric temperature is the measure of global warming often reported in the popular press, most of the additional energy stored in the climate system since 1970 has gone into the oceans. The rest has melted ice and warmed the continents and atmosphere.

The average temperature of the lower troposphere has increased between 0.12 and 0.135 °C (0.216 and 0.243 °F) per decade since 1979, according to satellite temperature measurements. Climate proxies show the temperature to have been relatively stable over the one or two thousand years before 1850, with regionally varying fluctuations such as the Medieval Warm Period and the Little Ice Age.

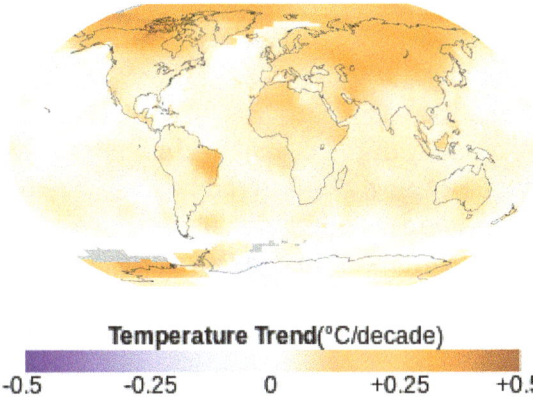

World map showing surface temperature trends (°C per decade) between 1950 and 2014.

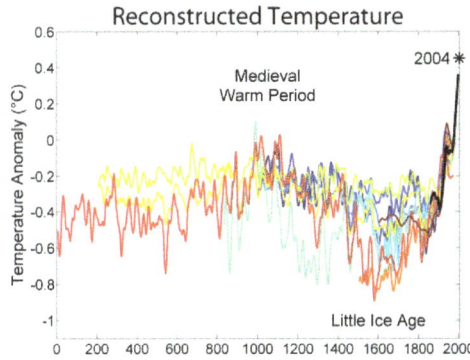

Two millennia of mean surface temperatures according to different reconstructions from climate proxies, each smoothed on a decadal scale, with the instrumental temperature record overlaid in black.

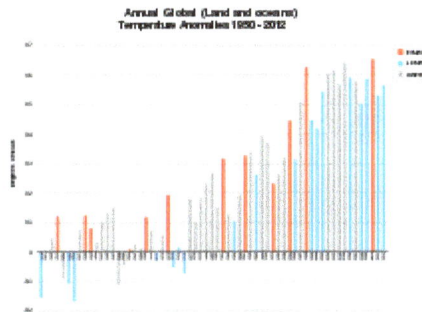

NOAA graph of global annual temperature anomalies 1950–2012, showing the El Niño Southern Oscillation.

The warming that is evident in the instrumental temperature record is consistent with a wide range of observations, as documented by many independent scientific groups. Examples include sea level rise, widespread melting of snow and land ice, increased heat content of the oceans, increased humidity, and the earlier timing of spring events, e.g., the flowering of plants. The probability that these changes could have occurred by chance is virtually zero.

Trends

Temperature changes vary over the globe. Since 1979, land temperatures have increased about twice as fast as ocean temperatures (0.25 °C per decade against 0.13 °C per decade). Ocean temperatures increase more slowly than land temperatures because of the larger effective heat capacity of the oceans and because the ocean loses more heat by evaporation. Since the beginning of industrialisation the temperature difference between the hemispheres has increased due to melting of sea ice and snow in the North. Average arctic temperatures have been increasing at almost twice the rate of the rest of the world in the past 100 years; however arctic temperatures are also highly variable. Although more greenhouse gases are emitted in the Northern than Southern Hemisphere this does not contribute to the difference in warming because the major greenhouse gases persist long enough to mix between hemispheres.

The thermal inertia of the oceans and slow responses of other indirect effects mean that climate can take centuries or longer to adjust to changes in forcing. One climate commitment study concluded that if greenhouse gases were stabilized at year 2000 levels, surface temperatures would still increase by about one-half degree Celsius, and another found that if they were stabilized at 2005 levels surface warming could exceed a whole degree Celsius. Some of this surface warming will be driven by past natural forcings which are still seeking equilibrium in the climate system. One study using a highly simplified climate model indicates these past natural forcings may account for as much as 64% of the committed 2050 surface warming and their influence will fade with time compared to the human contribution.

Global temperature is subject to short-term fluctuations that overlay long-term trends and can temporarily mask them. The relative stability in surface temperature from 2002 to 2009, which has been dubbed the global warming hiatus by the media and some scientists, is consistent with such an episode. 2015 updates to account for differing methods of measuring ocean surface temperature measurements show a positive trend over the recent decade.

Warmest Years

Sixteen of the 17 warmest years on record have occurred since 2000. While record-breaking years attract considerable public interest, individual years are less significant than the overall trend. Some climatologists have criticized the attention that the popular press gives to "warmest year" statistics. In particular, ocean oscillations such as the El Niño Southern Oscillation (ENSO) can cause temperatures of a given year to be abnormally warm or cold for reasons unrelated to the overall trend of climate change. Gavin Schmidt stated "the long-term trends or the expected sequence of records are far more important than whether any single year is a record or not."

Initial Causes of Temperature Changes (External Forcings)

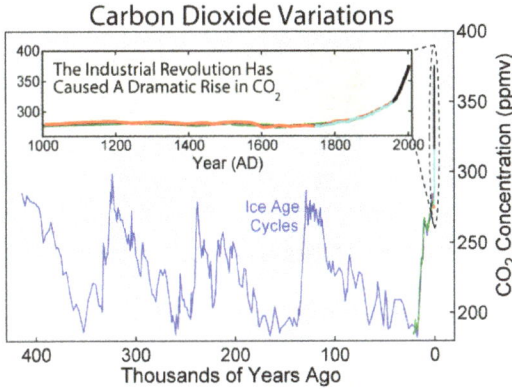

CO$_2$ concentrations over the last 400,000 years.

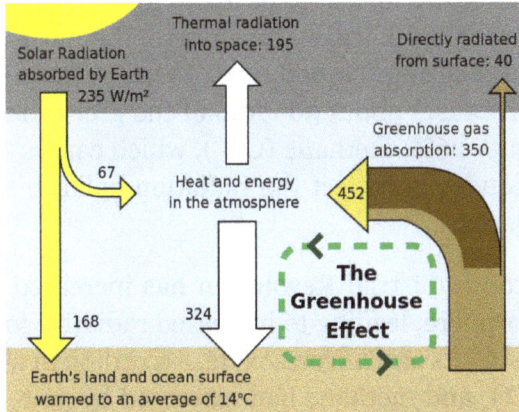

Greenhouse effect schematic showing energy flows between space, the atmosphere, and Earth's surface. Energy exchanges are expressed in watts per square metre (W/m²).

The climate system can spontaneously generate changes in global temperature for years to decades at a time but long-term changes in global temperature require *external forcings*. These forcings are "external" to the climate system but not necessarily external to Earth. Examples of external forcings include changes in atmospheric composition (e.g., increased concentrations of greenhouse gases), solar luminosity, volcanic eruptions, and variations in Earth's orbit around the Sun.

Greenhouse Gases

The greenhouse effect is the process by which absorption and emission of infrared radiation by gases in a planet's atmosphere warm its lower atmosphere and surface. It was proposed by Joseph Fourier in 1824, discovered in 1860 by John Tyndall, was first investigated quantitatively by Svante Arrhenius in 1896, and its scientific description was developed in the 1930s through 1960s by Guy Stewart Callendar.

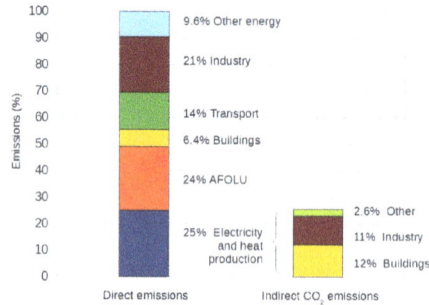

Annual world greenhouse gas emissions, in 2010, by sector.

Percentage share of global cumulative energy-related CO_2 emissions between 1751 and 2012 across different regions.

On Earth, an atmosphere containing naturally occurring amounts of greenhouse gases causes air temperature near the surface to be about 33 °C (59 °F) warmer than it would be in their absence. Without the Earth's atmosphere, the Earth's average temperature would be well below the freezing temperature of water. The major greenhouse gases are water vapour, which causes about 36–70% of the greenhouse effect; carbon dioxide (CO_2), which causes 9–26%; methane (CH_4), which causes 4–9%; and ozone (O_3), which causes 3–7%. Clouds also affect the radiation balance through cloud forcings similar to greenhouse gases.

Human activity since the Industrial Revolution has increased the amount of greenhouse gases in the atmosphere, leading to increased radiative forcing from CO_2, methane, tropospheric ozone, CFCs and nitrous oxide. According to work published in 2007, the concentrations of CO_2 and methane had increased by 36% and 148% respectively since 1750. These levels are much higher than at any time during the last 800,000 years, the period for which reliable data has been extracted from ice cores. Less direct geological evidence indicates that CO_2 values higher than this were last seen about 20 million years ago.

Fossil fuel burning has produced about three-quarters of the increase in CO_2 from human activity over the past 20 years. The rest of this increase is caused mostly by changes in land-use, particularly deforestation. Another significant non-fuel source of anthropogenic CO_2 emissions is the calcination of limestone for clinker production, a chemical process which releases CO_2. Estimates of global CO_2 emissions in 2011 from fossil fuel combustion, including cement production and gas flaring, was 34.8 billion tonnes (9.5 ± 0.5 PgC), an increase of 54% above emissions in 1990. Coal burning was responsible for 43% of the total emissions, oil 34%, gas 18%, cement 4.9% and gas flaring 0.7%.

In May 2013, it was reported that readings for CO_2 taken at the world's primary benchmark site in Mauna Loa surpassed 400 ppm. According to professor Brian Hoskins, this is likely the first time CO_2 levels have been this high for about 4.5 million years. Monthly global CO_2 concentrations exceeded 400 ppm in March 2015, probably for the first time in several million years. On 12 November 2015, NASA scientists reported that human-made carbon dioxide continues to increase above levels not seen in hundreds of thousands of years: currently, about half of the carbon dioxide released from the burning of fossil fuels is not absorbed by vegetation and the oceans and remains in the atmosphere.

Over the last three decades of the twentieth century, gross domestic product per capita and population growth were the main drivers of increases in greenhouse gas emissions. CO_2 emissions are continuing to rise due to the burning of fossil fuels and land-use change. Emissions can be attributed to different regions. Attributions of emissions due to land-use change are subject to considerable uncertainty.

Emissions scenarios, estimates of changes in future emission levels of greenhouse gases, have been projected that depend upon uncertain economic, sociological, technological, and natural developments. In most scenarios, emissions continue to rise over the century, while in a few, emissions are reduced. Fossil fuel reserves are abundant, and will not limit carbon emissions in the 21st century. Emission scenarios, combined with modelling of the carbon cycle, have been used to produce estimates of how atmospheric concentrations of greenhouse gases might change in the future. Using the six IPCC SRES "marker" scenarios, models suggest that by the year 2100, the atmospheric concentration of CO_2 could range between 541 and 970 ppm. This is 90–250% above the concentration in the year 1750.

The popular media and the public often confuse global warming with ozone depletion, i.e., the destruction of stratospheric ozone (e.g., the ozone layer) by chlorofluorocarbons. Although there are a few areas of linkage, the relationship between the two is not strong. Reduced stratospheric ozone has had a slight cooling influence on surface temperatures, while increased tropospheric ozone has had a somewhat larger warming effect.

Aerosols and Soot

Global dimming, a gradual reduction in the amount of global direct irradiance at the Earth's surface, was observed from 1961 until at least 1990. Solid and liquid particles known as *aerosols*, produced by volcanoes and human-made pollutants, are thought to be the main cause of this dimming. They exert a cooling effect by increasing the reflection of incoming sunlight. The effects of the products of fossil fuel combustion – CO_2 and aerosols – have partially offset one another in recent decades, so that net warming has been due to the increase in non-CO_2 greenhouse gases such as methane. Radiative forcing due to aerosols is temporally limited due to the processes that remove aerosols from the atmosphere. Removal by clouds and precipitation gives tropospheric aerosols an atmospheric lifetime of only about a week, while stratospheric aerosols can remain for a few years. Carbon dioxide has a lifetime of a century or more, and as such, changes in aerosols will only delay climate changes due to carbon dioxide. Black carbon is second only to carbon dioxide for its contribution to global warming (contribution being estimated at 17 to 20%, whereas carbon dioxide contributes 40 to 45% to global warming).

Ship tracks can be seen as lines in these clouds over the Atlantic Ocean on the east coast of the United States. Atmospheric particles from these and other sources could have a large effect on climate through the aerosol indirect effect.

In addition to their direct effect by scattering and absorbing solar radiation, aerosols have indirect effects on the Earth's radiation budget. Sulfate aerosols act as cloud condensation nuclei and thus lead to clouds that have more and smaller cloud droplets. These clouds reflect solar radiation more efficiently than clouds with fewer and larger droplets, a phenomenon known as the Twomey effect. This effect also causes droplets to be of more uniform size, which reduces growth of raindrops and makes the cloud more reflective to incoming sunlight, known as the Albrecht effect. Indirect effects are most noticeable in marine stratiform clouds, and have very little radiative effect on convective clouds. Indirect effects of aerosols represent the largest uncertainty in radiative forcing.

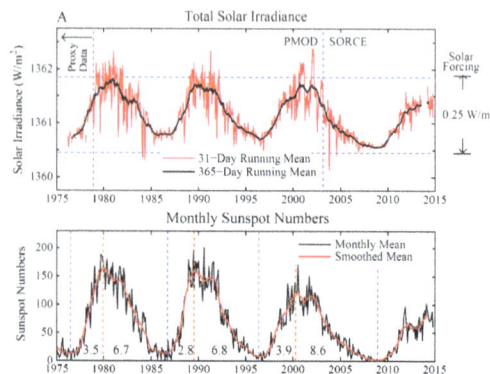

Changes in total solar irradiance (TSI) and monthly sunspot numbers since the mid-1970s.

Soot may either cool or warm Earth's climate system, depending on whether it is airborne or deposited. Atmospheric soot directly absorbs solar radiation, which heats the atmosphere and cools the surface. In isolated areas with high soot production, such as rural India, as much as 50% of surface warming due to greenhouse gases may be masked by atmospheric brown clouds. When deposited, especially on glaciers or on ice in arctic regions, the lower surface albedo can also directly heat the surface. The influences of atmospheric particles, including black carbon, are most pronounced in the tropics and sub-tropics, particularly in Asia, while the effects of greenhouse gases are dominant in the extratropics and southern hemisphere.

Radiative forcing components

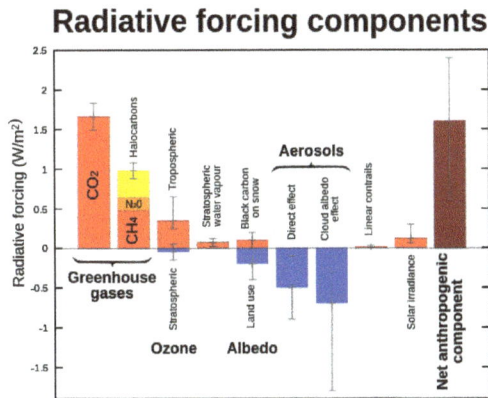

Contribution of natural factors and human activities to radiative forcing of climate change. Radiative forcing values are for the year 2005, relative to the pre-industrial era (1750). The contribution of solar irradiance to radiative forcing is 5% the value of the combined radiative forcing due to increases in the atmospheric concentrations of carbon dioxide, methane and nitrous oxide.

Solar Activity

Since 1978, solar irradiance has been measured by satellites. These measurements indicate that the Sun's radiative output has not increased during that time, so the warming during the past 40 years cannot be attributed to an increase in solar energy reaching the Earth.

Climate models have been used to examine the role of the Sun in recent climate change. Models are unable to reproduce the rapid warming observed in recent decades when they only take into account variations in solar output and volcanic activity. Models are, however, able to simulate the observed 20th century changes in temperature when they include all of the most important external forcings, including human influences and natural forcings.

Another line of evidence is differing temperature changes at different levels in the Earth's atmosphere. Basic physical principles require that the greenhouse effect produces warming of the lower atmosphere (the troposphere) but cooling of the upper atmosphere (the stratosphere). Depletion of the ozone layer by chemical refrigerants has also resulted in a strong cooling effect in the stratosphere. If solar variations were responsible for observed warming, warming of both the troposphere and stratosphere would be expected.

Variations in Earth's Orbit

The tilt of the Earth's axis and the shape of its orbit around the Sun vary slowly over tens of thousands of years. This changes climate by changing the seasonal and latitudinal distribution of incoming solar energy at Earth's surface. During the last few thousand years, this phenomenon contributed to a slow cooling trend at high latitudes of the Northern Hemisphere during summer, a trend that was reversed by greenhouse-gas-induced warming during the 20th century. Orbital cycles favorable for glaciation are not expected within the next 50,000 years.

Feedback

Sea ice, shown here in Nunavut, in northern Canada, reflects more sunshine, while open ocean absorbs more, accelerating melting.

The climate system includes a range of *feedbacks*, which alter the response of the system to changes in external forcings. Positive feedbacks increase the response of the climate system to an initial forcing, while negative feedbacks reduce it.

There are a range of feedbacks in the climate system, including water vapour, changes in ice-albedo (snow and ice cover affect how much the Earth's surface absorbs or reflects incoming sunlight), clouds, and changes in the Earth's carbon cycle (e.g., the release of carbon from soil). The main negative feedback is the energy the Earth's surface radiates into space as infrared radiation. According to the Stefan-Boltzmann law, if the absolute temperature (as measured in kelvins) doubles, radiated energy increases by a factor of 16 (2 to the 4th power).

Feedbacks are an important factor in determining the sensitivity of the climate system to increased atmospheric greenhouse gas concentrations. Other factors being equal, a higher *climate sensitivity* means that more warming will occur for a given increase in greenhouse gas forcing. Uncertainty over the effect of feedbacks is a major reason why different climate models project different magnitudes of warming for a given forcing scenario. More research is needed to understand the role of clouds and carbon cycle feedbacks in climate projections.

The IPCC projections previously mentioned span the "likely" range (greater than 66% probability, based on expert judgement) for the selected emissions scenarios. However, the IPCC's projections do not reflect the full range of uncertainty. The lower end of the "likely" range appears to be better constrained than the upper end.

Climate Models

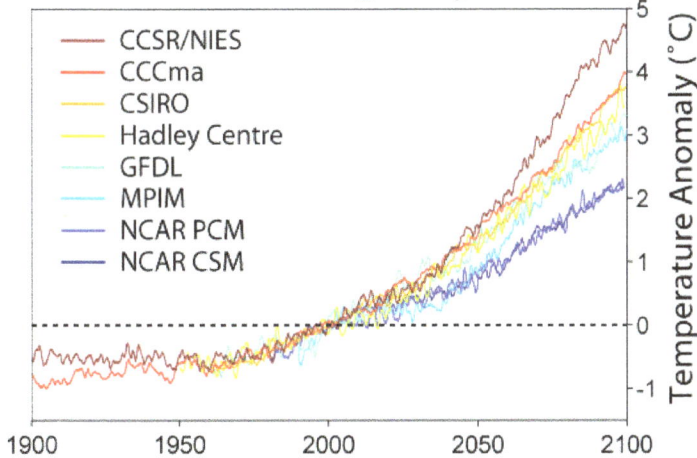

Global Warming Projections

Legend:
— CCSR/NIES
— CCCma
— CSIRO
— Hadley Centre
— GFDL
— MPIM
— NCAR PCM
— NCAR CSM

Calculations of global warming prepared in or before 2001 from a range of climate models under the SRES A2 emissions scenario, which assumes no action is taken to reduce emissions and regionally divided economic development.

A climate model is a representation of the physical, chemical and biological processes that affect the climate system. Such models are based on scientific disciplines such as fluid dynamics and thermodynamics as well as physical processes such as radiative transfer. The models may be used to predict a range of variables such as local air movement, temperature, clouds, and other atmospheric properties; ocean temperature, salt content, and circulation; ice cover on land and sea; the transfer of heat and moisture from soil and vegetation to the atmosphere; and chemical and biological processes, among others.

Although researchers attempt to include as many processes as possible, simplifications of the actual climate system are inevitable because of the constraints of available computer power and limitations in knowledge of the climate system. Results from models can also vary due to different greenhouse gas inputs and the model's climate sensitivity. For example, the uncertainty in IPCC's 2007 projections is caused by (1) the use of multiple models with differing sensitivity to greenhouse gas concentrations, (2) the use of differing estimates of humanity's future greenhouse gas emissions, (3) any additional emissions from climate feedbacks that were not included in the models IPCC used to prepare its report, i.e., greenhouse gas releases from permafrost.

The models do not assume the climate will warm due to increasing levels of greenhouse gases. Instead the models predict how greenhouse gases will interact with radiative transfer and other physical processes. Warming or cooling is thus a result, not an assumption, of the models.

Clouds and their effects are especially difficult to predict. Improving the models' representation of clouds is therefore an important topic in current research. Another prominent research topic is expanding and improving representations of the carbon cycle.

Models are also used to help investigate the causes of recent climate change by comparing the observed changes to those that the models project from various natural and human causes. Although these models do not unambiguously attribute the warming that occurred from approximately 1910 to 1945 to either natural variation or human effects, they do indicate that the warming since 1970 is dominated by anthropogenic greenhouse gas emissions.

The physical realism of models is tested by examining their ability to simulate contemporary or past climates. Climate models produce a good match to observations of global temperature changes over the last century, but do not simulate all aspects of climate. Not all effects of global warming are accurately predicted by the climate models used by the IPCC. Observed Arctic shrinkage has been faster than that predicted. Precipitation increased proportionally to atmospheric humidity, and hence significantly faster than global climate models predict. Since 1990, sea level has also risen considerably faster than models predicted it would.

Observed and Expected Environmental Effects

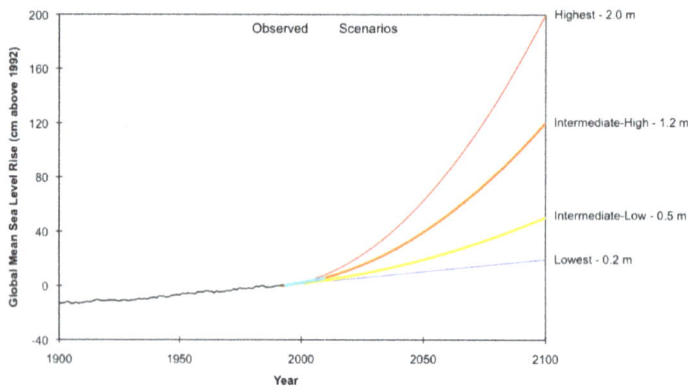

Projections of global mean sea level rise by Parris and others. Probabilities have not been assigned to these projections. Therefore, none of these projections should be interpreted as a "best estimate" of future sea level rise.

Anthropogenic forcing has likely contributed to some of the observed changes, including sea level rise, changes in climate extremes (such as the number of warm and cold days), declines in Arctic sea ice extent, glacier retreat, and greening of the Sahara.

During the 21st century, glaciers and snow cover are projected to continue their widespread retreat. Projections of declines in Arctic sea ice vary. Recent projections suggest that Arctic summers could be ice-free (defined as ice extent less than 1 million square km) as early as 2025–2030.

"Detection" is the process of demonstrating that climate has changed in some defined statistical sense, without providing a reason for that change. Detection does not imply attribution of the detected change to a particular cause. "Attribution" of causes of climate change is the process of establishing the most likely causes for the detected change with some defined level of confidence. Detection and attribution may also be applied to observed changes in physical, ecological and social systems.

Extreme Weather

Changes in regional climate are expected to include greater warming over land, with most warming at high northern latitudes, and least warming over the Southern Ocean and parts of the North Atlantic Ocean.

Future changes in precipitation are expected to follow existing trends, with reduced precipitation over subtropical land areas, and increased precipitation at subpolar latitudes and some equatorial regions. Projections suggest a probable increase in the frequency and severity of some extreme weather events, such as heat waves.

A 2015 study published in *Nature Climate Change*, states:

About 18% of the moderate daily precipitation extremes over land are attributable to the observed temperature increase since pre-industrial times, which in turn primarily results from human influence. For 2 °C of warming the fraction of precipitation extremes attributable to human influence rises to about 40%. Likewise, today about 75% of the moderate daily hot extremes over land are attributable to warming. It is the most rare and extreme events for which the largest fraction is anthropogenic, and that contribution increases nonlinearly with further warming.

Data analysis of extreme events from 1960 until 2010 suggests that droughts and heat waves appear simultaneously with increased frequency. Extremely wet or dry events within the monsoon period have increased since 1980.

Sea Level Rise

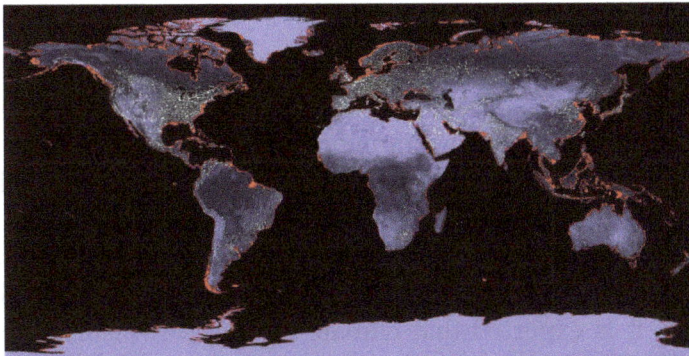

Map of the Earth with a six-metre sea level rise represented in red.

The sea level rise since 1993 has been estimated to have been on average 2.6 mm and 2.9 mm per year ± 0.4 mm. Additionally, sea level rise has accelerated from 1995 to 2015. Over the 21st century, the IPCC projects for a high emissions scenario, that global mean sea level could rise by 52–98 cm. The IPCC's projections are conservative, and may underestimate future sea level rise. Other estimates suggest that for the same period, global mean sea level could rise by 0.2 to 2.0 m (0.7–6.6 ft), relative to mean sea level in 1992.

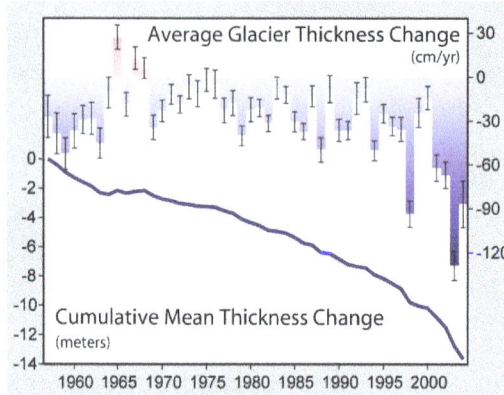

Sparse records indicate that glaciers have been retreating since the early 1800s. In the 1950s measurements began that allow the monitoring of glacial mass balance, reported to the World Glacier Monitoring Service (WGMS) and the National Snow and Ice Data Center (NSIDC).

Widespread coastal flooding would be expected if several degrees of warming is sustained for millennia. For example, sustained global warming of more than 2 °C (relative to pre-industrial levels) could lead to eventual sea level rise of around 1 to 4 m due to thermal expansion of sea water and the melting of glaciers and small ice caps. Melting of the Greenland ice sheet could contribute an additional 4 to 7.5 m over many thousands of years. It has been estimated that we are already committed to a sea-level rise of approximately 2.3 metres for each degree of temperature rise within the next 2,000 years.

Warming beyond the 2 °C target would potentially lead to rates of sea-level rise dominated by ice loss from Antarctica. Continued CO_2 emissions from fossil sources could cause additional tens of metres of sea level rise, over the next millennia and eventually ultimately eliminate the entire Antarctic ice sheet, causing about 58 metres of sea level rise.

Ecological Systems

In terrestrial ecosystems, the earlier timing of spring events, as well as poleward and upward shifts in plant and animal ranges, have been linked with high confidence to recent warming. Future climate change is expected to affect particular ecosystems, including tundra, mangroves, coral reefs, and caves. It is expected that most ecosystems will be affected by higher atmospheric CO_2 levels, combined with higher global temperatures. Overall, it is expected that climate change will result in the extinction of many species and reduced diversity of ecosystems.

Increases in atmospheric CO_2 concentrations have led to an increase in ocean acidity. Dissolved CO_2 increases ocean acidity, measured by lower pH values. Between 1750 and 2000, surface-ocean pH has decreased by ≈ 0.1, from ≈ 8.2 to ≈ 8.1. Surface-ocean pH has probably not been below ≈ 8.1 during the past 2 million years. Projections suggest that surface-ocean pH could decrease by an additional 0.3–0.4 units by 2100. Future ocean acidification could threaten coral reefs, fisheries, protected species, and other natural resources of value to society.

Ocean deoxygenation is projected to increase hypoxia by 10%, and triple suboxic waters (oxygen concentrations 98% less than the mean surface concentrations), for each 1 °C of upper ocean warming.

Long-term Effects

On the timescale of centuries to millennia, the magnitude of global warming will be determined primarily by anthropogenic CO_2 emissions. This is due to carbon dioxide's very long lifetime in the atmosphere.

Stabilizing the global average temperature would require large reductions in CO_2 emissions, as well as reductions in emissions of other greenhouse gases such as methane and nitrous oxide. Emissions of CO_2 would need to be reduced by more than 80% relative to their peak level. Even if this were achieved, global average temperatures would remain close to their highest level for many centuries. As of 2016, emissions of CO_2 from burning fossil fuels had stopped increasing, but the Guardian reports they need to be "reduced to have a real impact on climate change". Meanwhile, this greenhouse gas continues to accumulate in the atmosphere. Also, CO_2 is not the only factor driving climate change. Concentrations of atmospheric methane, another greenhouse gas, rose dramatically between 2006–2016 for unknown reasons. This undermines efforts to combat global warming and there is a risk of an uncontrollable runaway greenhouse effect.

Long-term effects also include a response from the Earth's crust, due to ice melting and deglaciation, in a process called post-glacial rebound, when land masses are no longer depressed by the weight of ice. This could lead to landslides and increased seismic and volcanic activities. Tsunamis could be generated by submarine landslides caused by warmer ocean water thawing ocean-floor permafrost or releasing gas hydrates. Some world regions, such as the French Alps, already show signs of an increase in landslide frequency.

Large-scale and Abrupt Impacts

Climate change could result in global, large-scale changes in natural and social systems. Examples include the possibility for the Atlantic Meridional Overturning Circulation to slow- or shutdown, which in the instance of a shutdown would change weather in Europe and North America considerably, ocean acidification caused by increased atmospheric concentrations of carbon dioxide, and the long-term melting of ice sheets, which contributes to sea level rise.

Some large-scale changes could occur abruptly, i.e., over a short time period, and might also be irreversible. Examples of abrupt climate change are the rapid release of methane and carbon dioxide from permafrost, which would lead to amplified global warming, or the shutdown of thermohaline circulation. Scientific understanding of abrupt climate change is generally poor. The probability of abrupt change for some climate related feedbacks may be low. Factors that may increase the probability of abrupt climate change include higher magnitudes of global warming, warming that occurs more rapidly, and warming that is sustained over longer time periods.

Observed and Expected Effects on Social Systems

The effects of climate change on human systems, mostly due to warming or shifts in precipitation patterns, or both, have been detected worldwide. Production of wheat and maize globally has been impacted by climate change. While crop production has increased in some mid-latitude regions such as the UK and Northeast China, economic losses due to extreme weather events have increased globally. There has been a shift from cold- to heat-related mortality in some regions as a result of warming. Livelihoods of indigenous peoples of the Arctic have been altered by climate change, and there is emerging evidence of climate change impacts on livelihoods of indigenous peoples in other regions. Regional impacts of climate change are now observable at more locations than before, on all continents and across ocean regions.

The future social impacts of climate change will be uneven. Many risks are expected to increase with higher magnitudes of global warming. All regions are at risk of experiencing negative impacts. Low-latitude, less developed areas face the greatest risk. A study from 2015 concluded that economic growth (gross domestic product) of poorer countries is much more impaired with projected future climate warming, than previously thought.

A meta-analysis of 56 studies concluded in 2014 that each degree of temperature rise will increase violence by up to 20%, which includes fist fights, violent crimes, civil unrest or wars.

Examples of impacts include:

- *Food*: Crop production will probably be negatively affected in low latitude countries, while effects at northern latitudes may be positive or negative. Global warming of around 4.6 °C relative to pre-industrial levels could pose a large risk to global and regional food security.

- *Health*: Generally impacts will be more negative than positive. Impacts include: the effects of extreme weather, leading to injury and loss of life; and indirect effects, such as undernutrition brought on by crop failures.

Habitat Inundation

In small islands and mega deltas, inundation as a result of sea level rise is expected to threaten vital infrastructure and human settlements. This could lead to issues of homelessness in countries with low-lying areas such as Bangladesh, as well as statelessness for populations in countries such as the Maldives and Tuvalu.

Economy

Estimates based on the IPCC A1B emission scenario from additional CO_2 and CH_4 greenhouse gases released from permafrost, estimate associated impact damages by US$43 trillion.

Infrastructure

Continued permafrost degradation will likely result in unstable infrastructure in Arctic regions, or Alaska before 2100. Thus, impacting roads, pipelines and buildings, as well as water distribution, and cause slope failures.

Possible Responses to Global Warming

Mitigation

The graph on the right shows three "pathways" to meet the UNFCCC's 2 °C target, labelled "global technology", "decentralized solutions", and "consumption change". Each pathway shows how various measures (e.g., improved energy efficiency, increased use of renewable energy) could contribute to emissions reductions. Image credit: PBL Netherlands Environmental Assessment Agency.

Mitigation of climate change are actions to reduce greenhouse gas emissions, or enhance the capacity of carbon sinks to absorb GHGs from the atmosphere. There is a large potential for future reductions in emissions by a combination of activities, including: energy conservation and increased energy efficiency; the use of low-carbon energy technologies, such as renewable energy, nuclear energy, and carbon capture and storage; and enhancing carbon sinks through, for example, reforestation and preventing deforestation. A 2015 report by Citibank concluded that transitioning to a low carbon economy would yield positive return on investments.

Near- and long-term trends in the global energy system are inconsistent with limiting global warming at below 1.5 or 2 °C, relative to pre-industrial levels. Pledges made as part of the Cancún agreements are broadly consistent with having a likely chance (66 to 100% probability) of limiting global warming (in the 21st century) at below 3 °C, relative to pre-industrial levels.

In limiting warming at below 2 °C, more stringent emission reductions in the near-term would allow for less rapid reductions after 2030. Many integrated models are unable to meet the 2 °C target if pessimistic assumptions are made about the availability of mitigation technologies.

Adaptation

Other policy responses include adaptation to climate change. Adaptation to climate change may be planned, either in reaction to or anticipation of climate change, or spontaneous, i.e., without government intervention. Planned adaptation is already occurring on a limited basis. The barriers, limits, and costs of future adaptation are not fully understood.

A concept related to adaptation is *adaptive capacity*, which is the ability of a system (human, natural or managed) to adjust to climate change (including climate variability and extremes) to moderate potential damages, to take advantage of opportunities, or to cope with consequences. Unmitigated climate change (i.e., future climate change without efforts to limit greenhouse gas emissions) would, in the long term, be likely to exceed the capacity of natural, managed and human systems to adapt.

Environmental organizations and public figures have emphasized changes in the climate and the risks they entail, while promoting adaptation to changes in infrastructural needs and emissions reductions.

Climate Engineering

Climate engineering (sometimes called *geoengineering* or *climate intervention*) is the deliberate modification of the climate. It has been investigated as a possible response to global warming, e.g. by NASA and the Royal Society. Techniques under research fall generally into the categories solar radiation management and carbon dioxide removal, although various other schemes have been suggested. A study from 2014 investigated the most common climate engineering methods and concluded they are either ineffective or have potentially severe side effects and cannot be stopped without causing rapid climate change.

Discourse about Global Warming

Political Discussion

Article 2 of the UN Framework Convention refers explicitly to "stabilization of greenhouse gas concentrations." To stabilize the atmospheric concentration of CO_2, emissions worldwide would need to be dramatically reduced from their present level.

Most countries in the world are parties to the United Nations Framework Convention on Climate Change (UNFCCC). The ultimate objective of the Convention is to prevent dangerous human interference of the climate system. As stated in the Convention, this requires that GHG concentrations are stabilized in the atmosphere at a level where ecosystems can adapt naturally to climate change, food production is not threatened, and economic development can proceed in a sustainable fashion. The Framework Convention was agreed in 1992, but since then, global emissions have risen.

During negotiations, the G77 (a lobbying group in the United Nations representing 133 developing nations) pushed for a mandate requiring developed countries to "[take] the lead" in reducing their emissions. This was justified on the basis that: the developed world's emissions had contributed most to the cumulation of GHGs in the atmosphere; per-capita emissions (i.e., emissions per head of population) were still relatively low in developing countries; and the emissions of developing countries would grow to meet their development needs.

This mandate was sustained in the Kyoto Protocol to the Framework Convention, which entered into legal effect in 2005. In ratifying the Kyoto Protocol, most developed countries accepted legally binding commitments to limit their emissions. These first-round commitments expired in 2012. United States President George W. Bush rejected the treaty on the basis that "it exempts 80% of the world, including major population centres such as China and India, from compliance, and would cause serious harm to the US economy."

At the 15th UNFCCC Conference of the Parties, held in 2009 at Copenhagen, several UNFCCC Parties produced the Copenhagen Accord. Parties associated with the Accord (140 countries, as of November 2010) aim to limit the future increase in global mean temperature to below 2 °C. The 16th Conference of the Parties (COP16) was held at Cancún in 2010. It produced an agreement, not a binding treaty, that the Parties should take urgent action to reduce greenhouse gas emissions to meet a goal of limiting global warming to 2 °C above pre-industrial temperatures. It also recognized the need to consider strengthening the goal to a global average rise of 1.5 °C.

Scientific Discussion

There is continuing discussion through published peer-reviewed scientific papers, which are assessed by scientists working in the relevant fields taking part in the Intergovernmental Panel on Climate Change. The scientific consensus as of 2013 stated in the IPCC Fifth Assessment Report is that it "is extremely likely that human influence has been the dominant cause of the observed warming since the mid-20th century". A 2008 report by the U.S. National Academy of Sciences stated that most scientists by then agreed that observed warming in recent decades was primarily caused by human activities increasing the amount of greenhouse gases in the atmosphere. In 2005 the Royal Society stated that while the overwhelming majority of scientists were in agreement on the main points, some individuals and organizations opposed to the consensus on urgent action needed to reduce greenhouse gas emissions had tried to undermine the science and work of the IPCC. National science academies have called on world leaders for policies to cut global emissions.

In the scientific literature, there is a strong consensus that global surface temperatures have increased in recent decades and that the trend is caused mainly by human-induced emissions of greenhouse gases. No scientific body of national or international standing disagrees with this view.

Discussion by the Public and in Popular Media

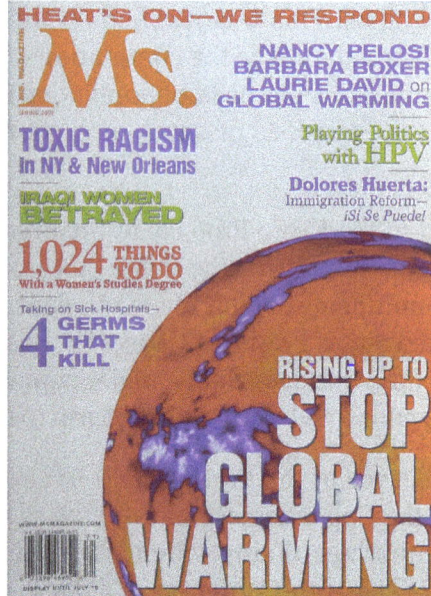

Global warming was the cover story in this 2007 issue of *Ms.* magazine.

The global warming controversy refers to a variety of disputes, substantially more pronounced in the popular media than in the scientific literature, regarding the nature, causes, and consequences of global warming. The disputed issues include the causes of increased global average air temperature, especially since the mid-20th century, whether this warming trend is unprecedented or within normal climatic variations, whether humankind has contributed significantly to it, and whether the increase is completely or partially an artefact of poor measurements. Additional disputes concern estimates of climate sensitivity, predictions of additional warming, and what the consequences of global warming will be.

By 1990, American conservative think tanks had begun challenging the legitimacy of global warming as a social problem. They challenged the scientific evidence, argued that global warming would have benefits, and asserted that proposed solutions would do more harm than good. Some people dispute aspects of climate change science. Organizations such as the libertarian Competitive Enterprise Institute, conservative commentators, and some companies such as ExxonMobil have challenged IPCC climate change scenarios, funded scientists who disagree with the scientific consensus, and provided their own projections of the economic cost of stricter controls. On the other hand, some fossil fuel companies have scaled back their efforts in recent years, or even called for policies to reduce global warming. Global oil companies have begun to acknowledge climate change exists and is caused by human activities and the burning of fossil fuels.

Surveys of Public Opinion

The world public, or at least people in economically advanced regions, became broadly aware of the global warming problem in the late 1980s. Polling groups began to track opinions on the subject, at first mainly in the United States. The longest consistent polling, by Gallup in the US, found relatively small deviations of 10% or so from 1998 to 2015 in opinion on the seriousness of global warming, but with increasing polarization between those concerned and those unconcerned.

The first major worldwide poll, conducted by Gallup in 2008–2009 in 127 countries, found that some 62% of people worldwide said they knew about global warming. In the advanced countries of North America, Europe and Japan, 90% or more knew about it (97% in the U.S., 99% in Japan); in less developed countries, especially in Africa, fewer than a quarter knew about it, although many had noticed local weather changes. Among those who knew about global warming, there was a wide variation between nations in belief that the warming was a result of human activities.

By 2010, with 111 countries surveyed, Gallup determined that there had been a substantial decrease since 2007–2008 in the number of Americans and Europeans who viewed global warming as a serious threat. In the US, just a little over half the population (53%) viewed it as a serious concern for either themselves or their families; this was 10 points below the 2008 poll (63%). Latin America had the biggest rise in concern: 73% said global warming was a serious threat to their families. This global poll also found that people were more likely to attribute global warming to human activities than to natural causes, except in the US where nearly half (47%) of the population attributed global warming to natural causes.

A March–May 2013 survey by Pew Research Center for the People & the Press polled 39 countries about global threats. According to 54% of those questioned, global warming featured top of the perceived global threats. In a January 2013 survey, Pew found that 69% of Americans say there is solid evidence that the Earth's average temperature has gotten warmer over the past few decades, up six points since November 2011 and 12 points since 2009.

A 2010 survey of 14 industrialized countries found that skepticism about the danger of global warming was highest in Australia, Norway, New Zealand and the United States, in that order, correlating positively with per capita emissions of carbon dioxide.

Etymology

In the 1950s, research suggested increasing temperatures, and a 1952 newspaper reported "climate change". This phrase next appeared in a November 1957 report in *The Hammond Times* which described Roger Revelle's research into the effects of increasing human-caused CO_2 emissions on the greenhouse effect, "a large scale global warming, with radical climate changes may result". Both phrases were only used occasionally until 1975, when Wallace Smith Broecker published a scientific paper on the topic,

"Climatic Change: Are We on the Brink of a Pronounced Global Warming?" The phrase began to come into common use, and in 1976 Mikhail Budyko's statement that "a global warming up has started" was widely reported. Other studies, such as a 1971 MIT report, referred to the human impact as "inadvertent climate modification", but an influential 1979 National Academy of Sciences study headed by Jule Charney followed Broecker in using *global warming* for rising surface temperatures, while describing the wider effects of increased CO_2 as *climate change.*

In 1986 and November 1987, NASA climate scientist James Hansen gave testimony to Congress on global warming. There were increasing heatwaves and drought problems in the summer of 1988, and when Hansen testified in the Senate on 23 June he sparked worldwide interest. He said: "global warming has reached a level such that we can ascribe with a high degree of confidence a cause and effect relationship between the greenhouse effect and the observed warming." Public attention increased over the summer, and *global warming* became the dominant popular term, commonly used both by the press and in public discourse.

In a 2008 NASA article on usage, Erik M. Conway defined *global warming* as "the increase in Earth's average surface temperature due to rising levels of greenhouse gases", while *climate change* was "a long-term change in the Earth's climate, or of a region on Earth." As effects such as changing patterns of rainfall and rising sea levels would probably have more impact than temperatures alone, he considered *global climate change* a more scientifically accurate term, and like the Intergovernmental Panel on Climate Change, the NASA website would emphasize this wider context.

Global Warming in Antarctica

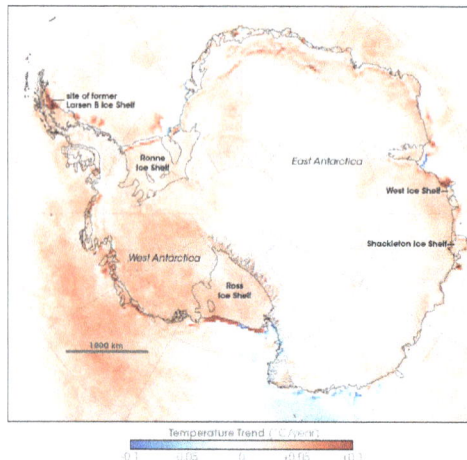

Antarctic Skin Temperature Trends between 1981 and 2007, based on thermal infrared observations made by a series of NOAA satellite sensors. Skin temperature trends do not necessarily reflect air temperature trends.

The effects of global warming in Antarctica may include rising temperatures and increasing snow melt.

Effects

The continent-wide average surface temperature trend of Antarctica is positive and significant at >0.05 °C/decade since 1957. The West Antarctic ice sheet has warmed by more than 0.1 °C/decade in the last 50 years, with most of the warming occurring in winter and spring. This is somewhat offset by cooling in East Antarctica during the fall. This effect is restricted to the 1980s and 1990s.

Research published in 2009 found that overall the continent had become warmer since the 1950s, a finding consistent with the influence of man-made climate change:

> "We can't pin it down, but it certainly is consistent with the influence of greenhouse gases from fossil fuels", said NASA scientist Drew Shindell, another study co-author. Some of the effects also could be natural variability, he said.

In 2017, a more detailed study found that the temperature trends had actually "shifted from a warming trend of 0.32 °C/decade during 1979-1997 to a cooling trend of −0.47 °C/decade during 1999-2014."

British Antarctic Survey

The British Antarctic Survey, which has undertaken the majority of Britain's scientific research in the area, has the following positions:

- Ice makes polar climate sensitive by introducing a strong positive feedback loop.

- Melting of continental Antarctic ice could contribute to global sea level rise.

- Climate models predict more snowfall than ice melting during the next 50 years, but the models are not good enough for them to be confident about the prediction.

- Antarctica seems to be both warming around the edges and cooling at the center at the same time. Thus it is not possible to say whether it is warming or cooling overall.

- There is no evidence for a decline in overall Antarctic sea ice extent.

- The central and southern parts of the west coast of the Antarctic Peninsula have warmed by about 2.4 °C. The cause is not known.

- Changes have occurred in the upper atmosphere over Antarctica.

The area of strongest cooling appears at the South Pole, and the region of strongest warming lies along the Antarctic Peninsula. A possible explanation is that loss of UV-absorbing ozone may have cooled the stratosphere and strengthened the polar vortex, a pattern of spinning winds around the South Pole. The vortex acts like an atmospheric barrier, preventing warmer, coastal air from moving into the continent's interior. A stronger polar vortex might explain the cooling trend in the interior of Antarctica.

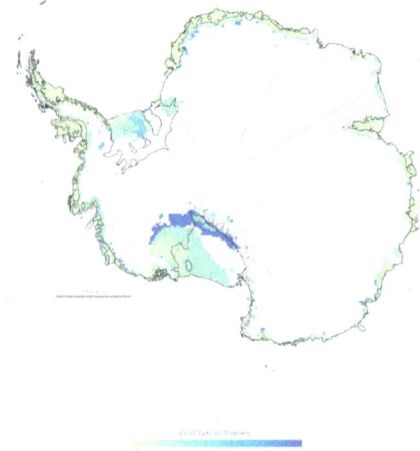

September 20, 2007 NASA map showing previously un-melted snowmelt.

In their latest study (September 20, 2007) NASA researchers have confirmed that Antarctic snow is melting farther inland from the coast over time, melting at higher altitudes than ever and increasingly melting on Antarctica's largest ice shelf.

There is also evidence for widespread glacier retreat around the Antarctic Peninsula.

Researchers reported December 21, 2012 in Nature Geoscience that from 1958 to 2010, the average temperature at the mile-high Byrd Station rose by 2.4 degrees Celsius, with warming fastest in its winter and spring. The spot which is in the heart of the West Antarctic Ice Sheet is one of the fastest-warming places on Earth.

Effects of Global Warming

The effects of global warming are the environmental and social changes caused (directly or indirectly) by human emissions of greenhouse gases. There is a scientific consensus that climate change is occurring, and that human activities are the primary driver. Many impacts of climate change have already been observed, including glacier retreat, changes in the timing of seasonal events (e.g., earlier flowering of plants), and changes in agricultural productivity.

Near-term climate change policies could significantly affect long-term climate change impacts. Stringent mitigation policies might be able to limit global warming (in 2100) to around 2 °C or below, relative to pre-industrial levels. Without mitigation, increased energy demand and extensive use of fossil fuels might lead to global warming of around 4 °C. Higher magnitudes of global warming would be more difficult to adapt to, and would increase the risk of negative impacts.

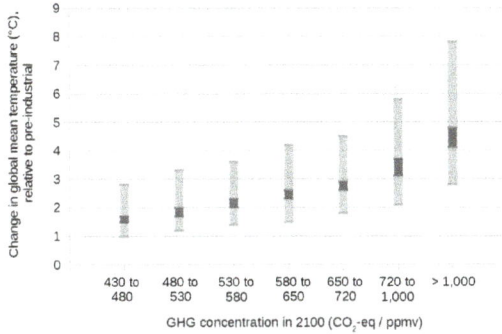

Projected global warming in 2100 for a range of emission scenarios.

Future effects of climate change will vary depending on climate change policies and social development. The two main policies to address climate change are reducing human greenhouse gas emissions (climate change mitigation) and adapting to the impacts of climate change. Geoengineering is another policy option.

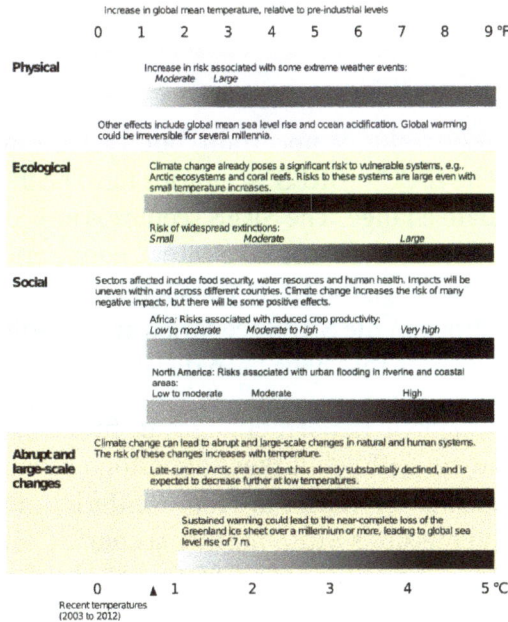

Summary of climate change impacts.

Definitions

"Climate change" means a change in climate that persists over a sustained period of time. The World Meteorological Organization defines this time period as 30 years. Examples of climate change include increases in global surface temperature (global warming), changes in rainfall patterns, and changes in the frequency of extreme weather events. Changes in climate may be due to natural causes, e.g., changes in the sun's output, or due to human activities, e.g., changing the composition of the atmosphere. Any human-induced changes in climate will occur against a background of natural climatic variations and of variations in human activity such as population growth on shores or in arid areas which increase or decrease climate vulnerability.

Also, the term "anthropogenic forcing" refers to the influence exerted on a habitat or chemical environment by humans, as opposed to a natural process.

Temperature Changes

The graph shows the average of a set of temperature simulations for the 20th century (black line), followed by projected temperatures for the 21st century based on three greenhouse gas emissions scenarios (colored lines).

Here is a description of the impacts of climate change according to different levels of future global warming. This way of describing impacts has, for instance, been used in the IPCC (Intergovernmental Panel on Climate Change) Assessment Reports on climate change. The instrumental temperature record shows global warming of around 0.6 °C during the 20th century.

SRES Emissions Scenarios

The future level of global warming is uncertain, but a wide range of estimates (projections) have been made. The IPCC's "SRES" scenarios have been frequently used to make projections of future climate change. The SRES scenarios are "baseline" (or "reference") scenarios, which means that they do not take into account any current or future measures to limit GHG emissions (e.g., the UNFCCC's Kyoto Protocol and the Cancún agreements). Emissions projections of the SRES scenarios are broadly comparable in range to the baseline emissions scenarios that have been developed by the scientific community.

In the IPCC Fourth Assessment Report, changes in future global mean temperature were projected using the six SRES "marker" emissions scenarios. Emissions projections for the six SRES "marker" scenarios are representative of the full set of forty SRES scenarios. For the lowest emissions SRES marker scenario ("B1"), the best estimate for global mean temperature is an increase of 1.8 °C (3.2 °F) by the end of the 21st century. This projection is relative to global temperatures at the end of the 20th century. The "likely" range (greater than 66% probability, based on expert judgement) for the SRES B1 marker scenario is 1.1–2.9 °C (2.0–5.2 °F). For the highest emissions SRES marker scenario (A1FI), the best estimate for global mean temperature increase is 4.0 °C (7.2 °F), with a "likely" range of 2.4–6.4 °C (4.3–11.5 °F).

The range in temperature projections partly reflects (1) the choice of emissions scenario, and (2) the "climate sensitivity". For (1), different scenarios make different assumptions of future social and economic development (e.g., economic growth, population level, energy policies), which in turn affects projections of greenhouse gas (GHG) emissions. The projected magnitude of warming by 2100 is closely related to the level of cumulative emissions over the 21st century (i.e. total emissions between 2000-2100). The higher the cumulative emissions over this time period, the greater the level of warming is projected to occur.

(2) reflects uncertainty in the response of the climate system to past and future GHG emissions, which is measured by the climate sensitivity). Higher estimates of climate sensitivity lead to greater projected warming, while lower estimates of climate sensitivity lead to less projected warming.

Over the next several millennia, projections suggest that global warming could be irreversible. Even if emissions were drastically reduced, global temperatures would remain close to their highest level for at least 1,000 years.

Projected Warming in Context

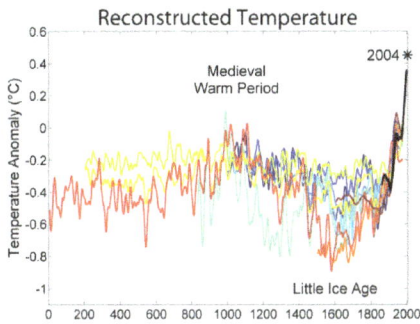

Two millennia of mean surface temperatures according to different reconstructions from climate proxies, each smoothed on a decadal scale, with the instrumental temperature record overlaid in black.

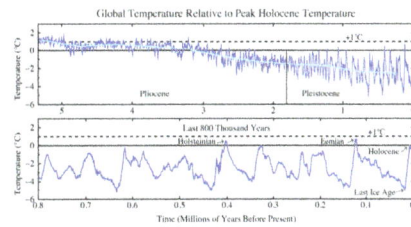

Global surface temperature for the past 5.3 million years as inferred from cores of ocean sediments taken all around the global ocean. The last 800,000 years are expanded in the lower half of the figure.

Scientists have used various "proxy" data to assess past changes in Earth's climate (paleoclimate). Sources of proxy data include historical records (such as farmers' logs), tree rings, corals, fossil pollen, ice cores, and ocean and lake sediments. Analysis of these data suggest that recent warming is unusual in the past 400 years, possibly longer. By the end of the 21st century, temperatures may increase to a level not experienced since the mid-Pliocene, around 3 million years ago. At that time, models suggest that mean global temperatures were about 2–3 °C warmer than pre-industrial temperatures. Even a 2 °C rise above the pre-industrial level would be outside the range of temperatures experienced by human civilization.

Physical Impacts

A broad range of evidence shows that the climate system has warmed. Evidence of global warming is shown in the graphs opposite. Some of the graphs show a positive trend, e.g., increasing temperature over land and the ocean, and sea level rise. Other graphs show a negative trend, e.g., decreased snow cover in the Northern Hemisphere, and declining Arctic sea ice extent. Evidence of warming is also apparent in living (biological) systems.

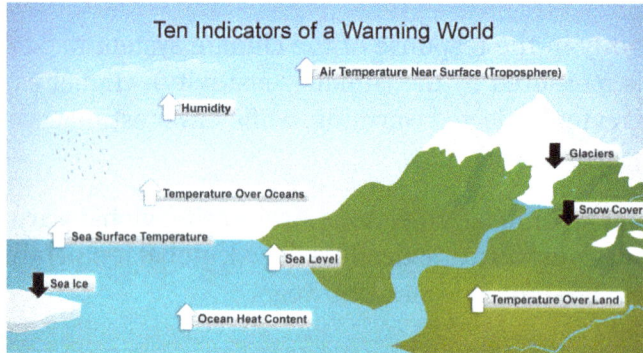

Seven of these indicators would be expected to increase in a warming world and observations show that they are, in fact, increasing. Three would be expected to decrease and they are, in fact, decreasing.

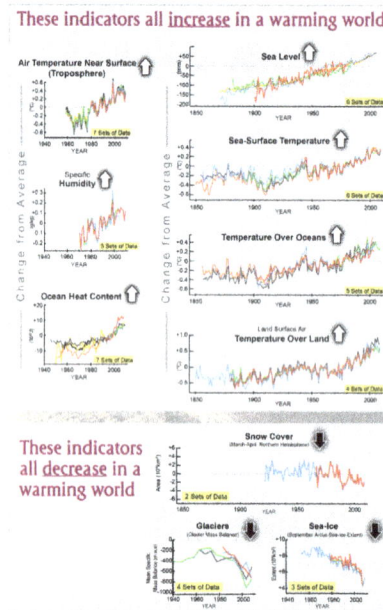

This set of graphs show changes in climate indicators over several decades. Each of the different colored lines in each panel represents an independently analyzed set of data. The data come from many different technologies including weather stations, satellites, weather balloons, ships and buoys.

Human activities have contributed to a number of the observed changes in climate. This contribution has principally been through the burning of fossil fuels, which has led to an increase in the concentration of GHGs in the atmosphere. Another human influence on the climate are sulfur dioxide emissions, which are a precursor to the formation of sulfate aerosols in the atmosphere.

Human-induced warming could lead to large-scale, irreversible, and/or abrupt changes in physical systems. An example of this is the melting of ice sheets, which contributes to sea level rise. The probability of warming having unforeseen consequences increases with the rate, magnitude, and duration of climate change.

Effects on Weather

Observations show that there have been changes in weather. As climate changes, the probabilities of certain types of weather events are affected.

Projected change in annual average precipitation by the end of the 21st century, based on a medium emissions scenario (SRES A1B).

Changes have been observed in the amount, intensity, frequency, and type of precipitation. Widespread increases in heavy precipitation have occurred, even in places where total rain amounts have decreased. With medium confidence, IPCC (2012) concluded that human influences had contributed to an increase in heavy precipitation events at the global scale.

Projections of future changes in precipitation show overall increases in the global average, but with substantial shifts in where and how precipitation falls. Projections suggest a reduction in rainfall in the subtropics, and an increase in precipitation in subpolar latitudes and some equatorial regions. In other words, regions which are dry at present will in general become even drier, while regions that are currently wet will in general become even wetter. This projection does not apply to every locale, and in some cases can be modified by local conditions.

Extreme Weather

Over most land areas since the 1950s, it is very likely that there have been fewer or warmer cold days and nights. Hot days and nights have also very likely become warmer or more frequent. Human activities have very likely contributed to these trends. There may have been changes in other climate extremes (e.g., floods, droughts and tropical cyclones) but these changes are more difficult to identify.

Projections suggest changes in the frequency and intensity of some extreme weather events. Confidence in projections varies over time.

Near-term projections (2016–2035)

Some changes (e.g., more frequent hot days) will probably be evident in the near term, while other near-term changes (e.g., more intense droughts and tropical cyclones) are more uncertain.

Long-term projections (2081–2100)

Future climate change will be associated with more very hot days and fewer very cold days. The frequency, length and intensity of heat waves will very likely increase over most land areas. Higher growth in anthropogenic GHG emissions will be associated with larger increases in the frequency and severity of temperature extremes.

Assuming high growth in GHG emissions (IPCC scenario RCP8.5), presently dry regions may be affected by an increase in the risk of drought and reductions in soil moisture. Over most of the mid-latitude land masses and wet tropical regions, extreme precipitation events will very likely become more intense and frequent.

Tropical cyclones

At the global scale, the frequency of tropical cyclones will probably decrease or be unchanged. Global mean tropical cyclone maximum wind speed and precipitation rates will likely increase. Changes in tropical cyclones will probably vary by region, but these variations are uncertain.

Effects of climate extremes

The impacts of extreme events on the environment and human society will vary. Some impacts will be beneficial—e.g., fewer cold extremes will probably lead to fewer cold deaths. Overall, however, impacts will probably be mostly negative.

Cryosphere

The cryosphere is made up of areas of the Earth which are covered by snow or ice. Observed changes in the cryosphere include declines in Arctic sea ice extent, the widespread retreat of alpine glaciers, and reduced snow cover in the Northern Hemisphere.

Solomon *et al.* (2007) assessed the potential impacts of climate change on summertime Arctic sea ice extent. Assuming high growth in greenhouse gas emissions (SRES A2), some models projected that Arctic sea ice in the summer could largely disappear by the end of the 21st century. More recent projections suggest that the Arctic summers could be ice-free (defined as ice extent less than 1 million square km) as early as 2025-2030.

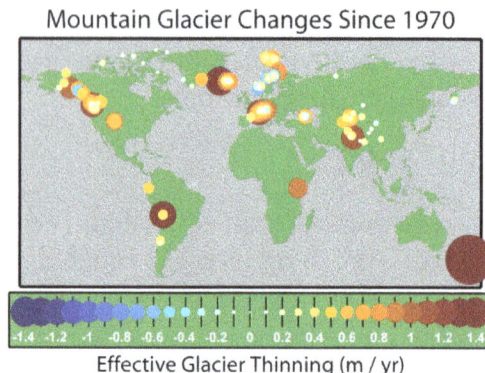

Mountain Glacier Changes Since 1970

Effective Glacier Thinning (m / yr)

A map of the change in thickness of mountain glaciers since 1970.
Thinning in orange and red, thickening in blue.

During the 21st century, glaciers and snow cover are projected to continue their wide-spread retreat. In the western mountains of North America, increasing temperatures and changes in precipitation are projected to lead to reduced snowpack. Snowpack is the seasonal accumulation of slow-melting snow. The melting of the Greenland and West Antarctic ice sheets could contribute to sea level rise, especially over long time-scales.

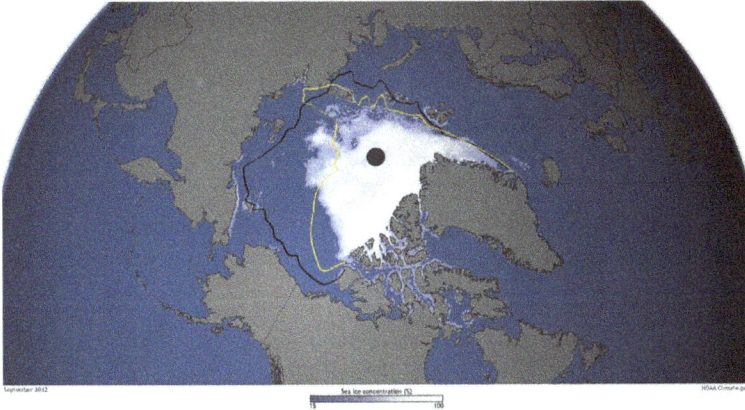

A map that shows ice concentration on 16 September 2012, along with the extent of the previous record low (yellow line) and the mid-September median extent (black line) setting a new record low that was 18 percent smaller than the previous record and nearly 50 percent smaller than the long-term (1979-2000) average.

Changes in the cryosphere are projected to have social impacts. For example, in some regions, glacier retreat could increase the risk of reductions in seasonal water availability. Barnett *et al.* (2005) estimated that more than one-sixth of the world's population rely on glaciers and snowpack for their water supply.

Oceans

The role of the oceans in global warming is complex. The oceans serve as a sink for carbon dioxide, taking up much that would otherwise remain in the atmosphere, but increased levels of CO_2 have led to ocean acidification. Furthermore, as the temperature of the oceans increases, they become less able to absorb excess CO_2. The ocean have also acted as a sink in absorbing extra heat from the atmosphere. The increase in ocean heat content is much larger than any other store of energy in the Earth's heat balance over the two periods 1961 to 2003 and 1993 to 2003, and accounts for more than 90% of the possible increase in heat content of the Earth system during these periods.

Global warming is projected to have a number of effects on the oceans. Ongoing effects include rising sea levels due to thermal expansion and melting of glaciers and ice sheets, and warming of the ocean surface, leading to increased temperature stratification. Other possible effects include large-scale changes in ocean circulation.

Acidification

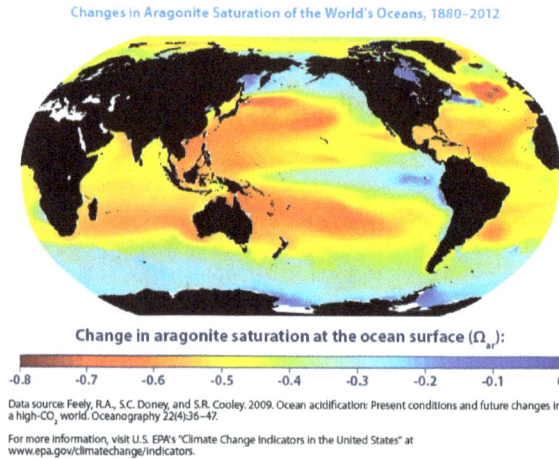

Changes in Aragonite Saturation of the World's Oceans, 1880–2012

Change in aragonite saturation at the ocean surface (Ω_{ar}):

-0.8   -0.7   -0.6   -0.5   -0.4   -0.3   -0.2   -0.1   0

Data source: Feely, R.A., S.C. Doney, and S.R. Cooley. 2009. Ocean acidification: Present conditions and future changes in a high-CO_2 world. Oceanography 22(4):36–47.

For more information, visit U.S. EPA's "Climate Change Indicators in the United States" at www.epa.gov/climatechange/indicators.

This map shows changes in the amount of aragonite dissolved in ocean surface waters between the 1880s and the most recent decade (2003-2012). Historical modeling suggests that since the 1880s, increased CO_2 has led to lower aragonite saturation levels (less availability of minerals) in the oceans around the world. The largest decreases in aragonite saturation have occurred in tropical waters. However, decreases in cold areas may be of greater concern because colder waters typically have lower aragonite levels to begin with.

About one-third of the carbon dioxide emitted by human activity has already been taken up by the oceans. As carbon dioxide dissolves in sea water, carbonic acid is formed, which has the effect of acidifying the ocean, measured as a change in pH. The uptake of human carbon emissions since the year 1750 has led to an average decrease in pH of 0.1 units. Projections using the SRES emissions scenarios suggest a further reduction in average global surface ocean pH of between 0.14 and 0.35 units over the 21st century.

The effects of ocean acidification on the marine biosphere have yet to be documented. Laboratory experiments suggest beneficial effects for a few species, with potentially highly detrimental effects for a substantial number of species. With medium confidence, Fischlin *et al.* (2007) projected that future ocean acidification and climate change would impair a wide range of planktonic and shallow benthic marine organisms that use aragonite to make their shells or skeletons, such as corals and marine snails (pteropods), with significant impacts particularly in the Southern Ocean.

Oxygen Depletion

The amount of oxygen dissolved in the oceans may decline, with adverse consequences for ocean life.

Sea Level Rise

There is strong evidence that global sea level rose gradually over the 20th century. With high confidence, Bindoff *et al.* (2007) concluded that between the mid-19th and mid-20th centuries, the rate of sea level rise increased. Authors of the IPCC Fourth Assessment Synthesis Report (IPCC AR4 SYR, 2007) reported that between the years 1961 and 2003, global average sea level rose at an average rate of 1.8 mm per year (mm/yr), with a range of 1.3–2.3 mm/yr. Between 1993 and 2003, the rate increased above the previous period to 3.1 mm/yr (range of 2.4–3.8 mm/yr). Authors of IPCC AR4 SYR (2007) were uncertain whether the increase in rate from 1993 to 2003 was due to natural variations in sea level over the time period, or whether it reflected an increase in the underlying long-term trend.

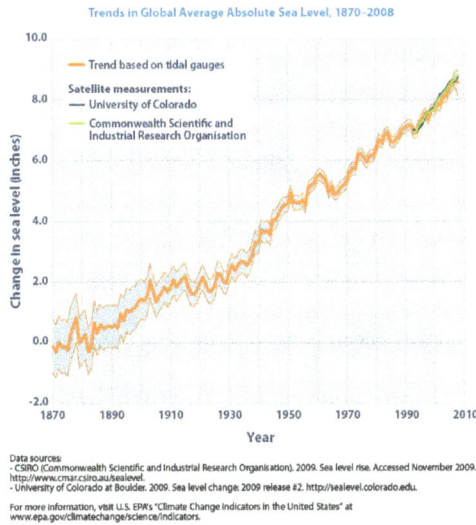

Trends in global average absolute sea level, 1870-2008.

There are two main factors that have contributed to observed sea level rise. The first is thermal expansion: as ocean water warms, it expands. The second is from the contribution of land-based ice due to increased melting. The major store of water on land is found in glaciers and ice sheets. Anthropogenic forces very likely (greater than 90% probability, based on expert judgement) contributed to sea level rise during the latter half of the 20th century.

There is a widespread consensus that substantial long-term sea level rise will continue for centuries to come. In their Fourth Assessment Report, the IPCC projected sea level rise to the end of the 21st century using the SRES emissions scenarios. Across the six SRES marker scenarios, sea level was projected to rise by 18 to 59 cm (7.1 to 23.2 in), relative to sea level at the end of the 20th century. Thermal expansion is the largest component in these projections, contributing 70-75% of the central estimate for all scenarios. Due to a lack of scientific understanding, this sea level rise estimate does not include all of the possible contributions of ice sheets.

An assessment of the scientific literature on climate change was published in 2010 by the US National Research Council (US NRC, 2010). NRC (2010) described the projections in AR4 (i.e. those cited in the above paragraph) as "conservative", and summarized the results of more recent studies. Cited studies suggested a great deal of uncertainty in projections. A range of projections suggested possible sea level rise by the end of the 21st century of between 0.56 and 2 m, relative to sea levels at the end of the 20th century.

Ocean Temperature Rise

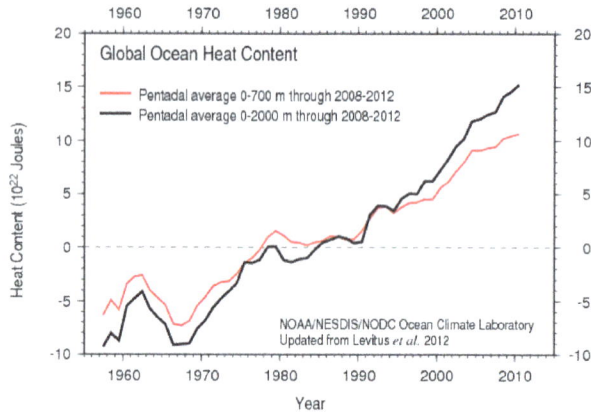

Global ocean heat content from 1955-2012

From 1961 to 2003, the global ocean temperature has risen by 0.10 °C from the surface to a depth of 700 m. There is variability both year-to-year and over longer time scales, with global ocean heat content observations showing high rates of warming for 1991–2003, but some cooling from 2003 to 2007. The temperature of the Antarctic Southern Ocean rose by 0.17 °C (0.31 °F) between the 1950s and the 1980s, nearly twice the rate for the world's oceans as a whole. As well as having effects on ecosystems (e.g. by melting sea ice, affecting algae that grow on its underside), warming reduces the ocean's ability to absorb CO 2. It is likely (greater than 66% probability, based on expert judgement) that anthropogenic forcing contributed to the general warming observed in the upper several hundred metres of the ocean during the latter half of the 20th century.

Regions

Temperatures across the world in the 1880s (left) and the 1980s (right), as compared to average temperatures from 1951 to 1980.

Projected changes in average temperatures across the world in the 2050s under three greenhouse gas (GHG) emissions scenarios.

Regional effects of global warming vary in nature. Some are the result of a generalised global change, such as rising temperature, resulting in local effects, such as melting ice. In other cases, a change may be related to a change in a particular ocean current or weather system. In such cases, the regional effect may be disproportionate and will not necessarily follow the global trend.

There are three major ways in which global warming will make changes to regional climate: melting or forming ice, changing the hydrological cycle (of evaporation and precipitation) and changing currents in the oceans and air flows in the atmosphere. The coast can also be considered a region, and will suffer severe impacts from sea level rise.

Observed Impacts

With very high confidence, Rosenzweig *et al.* (2007) concluded that physical and biological systems on all continents and in most oceans had been affected by recent climate changes, particularly regional temperature increases. Impacts include earlier leafing of trees and plants over many regions; movements of species to higher latitudes and altitudes in the Northern Hemisphere; changes in bird migrations in Europe, North America and Australia; and shifting of the oceans' plankton and fish from cold- to warm-adapted communities.

The human influence on the climate can be seen in the geographical pattern of observed warming, with greater temperature increases over land and in polar regions rather than over the oceans. Using models, it is possible to identify the human "signal" of global warming over both land and ocean areas.

Projected Impacts

Projections of future climate changes at the regional scale do not hold as high a level of scientific confidence as projections made at the global scale. It is, however, expected that future warming will follow a similar geographical pattern to that seen already, with greatest warming over land and high northern latitudes, and least over the Southern Ocean and parts of the North Atlantic Ocean. Nearly all land areas will very likely warm more than the global average.

The Arctic, Africa, small islands and Asian megadeltas are regions that are likely to be especially affected by climate change. Low-latitude, less-developed areas are at most risk of experiencing negative impacts due to climate change. Developed countries are also vulnerable to climate change. For example, developed countries will be negatively affected by increases in the severity and frequency of some extreme weather events, such as heat waves. In all regions, some people can be particularly at risk from climate change, such as the poor, young children and the elderly.

Social Systems

The impacts of climate change can be thought of in terms of sensitivity and vulnerability. "Sensitivity" is the degree to which a particular system or sector might be affected, positively or negatively, by climate change and/or climate variability. "Vulnerability" is the degree to which a particular system or sector might be adversely affected by climate change.

The sensitivity of human society to climate change varies. Sectors sensitive to climate change include water resources, coastal zones, human settlements, and human health. Industries sensitive to climate change include agriculture, fisheries, forestry, energy, construction, insurance, financial services, tourism, and recreation.

Food Supply

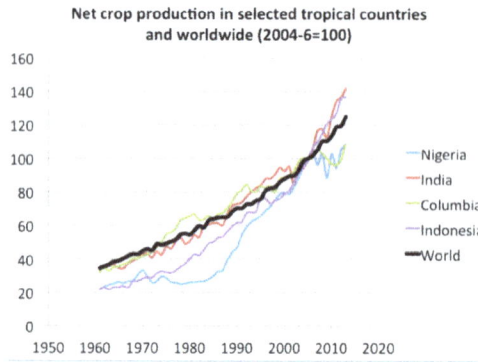

Graph of net crop production worldwide and in selected tropical countries.
Raw data from the United Nations.

Climate change will impact agriculture and food production around the world due to: the effects of elevated CO_2 in the atmosphere, higher temperatures, altered precipitation and transpiration regimes, increased frequency of extreme events, and modified weed, pest, and pathogen pressure. In general, low-latitude areas are at most risk of having decreased crop yields.

As of 2007, the effects of regional climate change on agriculture have been small. Changes in crop phenology provide important evidence of the response to recent regional climate change. Phenology is the study of natural phenomena that recur periodically, and how these phenomena relate to climate and seasonal changes. A significant advance in phenology has been observed for agriculture and forestry in large parts of the Northern Hemisphere.

Projections

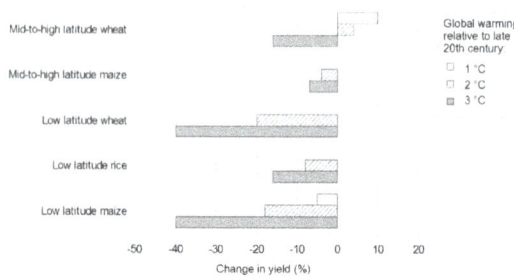

Projected changes in crop yields at different latitudes with global warming.
This graph is based on several studies.

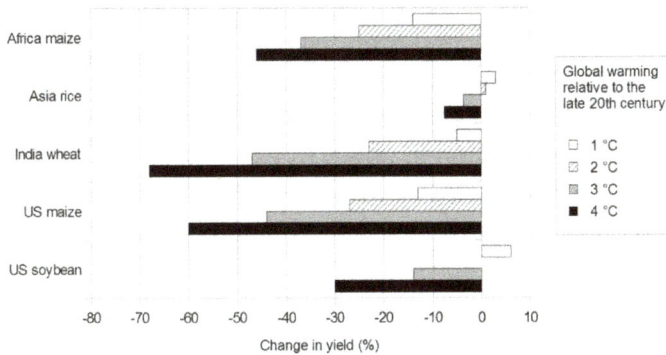

Projected changes in yields of selected crops with global warming.
This graph is based on several studies.

With low to medium confidence, Schneider *et al.* (2007) projected that for about a 1 to 3 °C increase in global mean temperature (by the years 2090-2100, relative to average temperatures in the years 1990–2000), there would be productivity decreases for some cereals in low latitudes, and productivity increases in high latitudes. With medium confidence, global production potential was projected to:

- increase up to around 3 °C,

- very likely decrease above about 3 °C.

Most of the studies on global agriculture assessed by Schneider *et al.* (2007) had not incorporated a number of critical factors, including changes in extreme events, or the spread of pests and diseases. Studies had also not considered the development of specific practices or technologies to aid adaptation to climate change.

The graphs opposite show the projected effects of climate change on selected crop yields. Actual changes in yields may be above or below these central estimates.

The projections above can be expressed relative to pre-industrial (1750) temperatures. 0.6 °C of warming is estimated to have occurred between 1750 and 1990-2000. Add 0.6 °C to the above projections to convert them from a 1990-2000 to pre-industrial baseline.

Food Security

Easterling *et al.* (2007) assessed studies that made quantitative projections of climate change impacts on food security. It was noted that these projections were highly uncertain and had limitations. However, the assessed studies suggested a number of fairly robust findings. The first was that climate change would likely increase the number of people at risk of hunger compared with reference scenarios with no climate change. Climate change impacts depended strongly on projected future social and economic development. Additionally, the magnitude of climate change impacts was projected to be smaller compared to the impact of social and economic development. In 2006, the global estimate for the number of people undernourished was 820 million. Under the SRES A1, B1, and B2 scenarios, projections for the year 2080 showed a reduction in the number of people undernourished of about 560-700 million people, with a global total of under-nourished people of 100-240 million in 2080. By contrast, the SRES A2 scenario showed only a small decrease in the risk of hunger from 2006 levels. The smaller reduction under A2 was attributed to the higher projected future population level in this scenario.

Droughts and Agriculture

Some evidence suggests that droughts have been occurring more frequently because of global warming and they are expected to become more frequent and intense in Africa, southern Europe, the Middle East, most of the Americas, Australia, and Southeast Asia. However, other research suggests that there has been little change in drought over the past 60 years. Their impacts are aggravated because of increased water demand, pop-ulation growth, urban expansion, and environmental protection efforts in many areas. Droughts result in crop failures and the loss of pasture grazing land for livestock.

Health

Human beings are exposed to climate change through changing weather patterns (tem-perature, precipitation, sea-level rise and more frequent extreme events) and indirectly through changes in water, air and food quality and changes in ecosystems, agriculture, industry and settlements and the economy (Confalonieri *et al.*, 2007:393). According to an assessment of the scientific literature by Confalonieri *et al.* (2007:393), the effects of climate change to date have been small, but are projected to progressively increase in all countries and regions.

A study by the World Health Organization (WHO, 2009) estimated the effect of climate change on human health. Not all of the effects of climate change were included in their estimates, for example, the effects of more frequent and extreme storms were excluded. Climate change was estimated to have been responsible for 3% of diarrhoea, 3% of ma-laria, and 3.8% of dengue fever deaths worldwide in 2004. Total attributable mortality was about 0.2% of deaths in 2004; of these, 85% were child deaths.

Projections

With high confidence, authors of the IPCC AR4 Synthesis report projected that climate change would bring some benefits in temperate areas, such as fewer deaths from cold exposure, and some mixed effects such as changes in range and transmission potential of malaria in Africa. Benefits were projected to be outweighed by negative health effects of rising temperatures, especially in developing countries.

With very high confidence, Confalonieri *et al.* (2007) concluded that economic develop-ment was an important component of possible adaptation to climate change. Economic growth on its own, however, was not judged to be sufficient to insulate the world's population from disease and injury due to climate change. Future vulnerability to climate change will depend not only on the extent of social and economic change, but also on how the benefits and costs of change are distributed in society. For example, in the 19th century, rapid urbanization in western Europe lead to a plummeting in population health. Other factors important in determining the health of populations include education, the availability of health services, and public-health infrastructure.

Water Resources

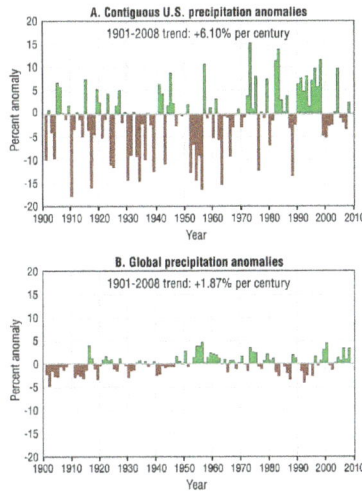

A. Contiguous U.S. precipitation anomalies
1901-2008 trend: +6.10% per century

B. Global precipitation anomalies
1901-2008 trend: +1.87% per century

[a]Anomalies and percent change are calculated with respect to the 1971-2000 mean.
Data source: NOAA, 2009

Precipitation during the 20th century and up through 2008 during global warming, the NOAA estimating an observed trend over that period of 1.87% global precipitation increase per century.

A number of climate-related trends have been observed that affect water resources. These include changes in precipitation, the crysosphere and surface waters (e.g., changes in river flows). Observed and projected impacts of climate change on freshwater systems and their management are mainly due to changes in temperature, sea level and precipitation variability. Sea level rise will extend areas of salinization of groundwater and estuaries, resulting in a decrease in freshwater availability for humans and ecosystems in coastal areas. In an assessment of the scientific literature, Kundzewicz *et al.* (2007) concluded, with high confidence, that:

- the negative impacts of climate change on freshwater systems outweigh the benefits. All of the regions assessed in the IPCC Fourth Assessment Report (Africa, Asia, Australia and New Zealand, Europe, Latin America, North America, Polar regions (Arctic and Antarctic), and small islands) showed an overall net negative impact of climate change on water resources and freshwater ecosystems.

- Semi-arid and arid areas are particularly exposed to the impacts of climate change on freshwater. With very high confidence, it was judged that many of these areas, e.g., the Mediterranean basin, Western United States, Southern Africa, and north-eastern Brazil, would suffer a decrease in water resources due to climate change.

Migration and Conflict

General circulation models project that the future climate change will bring wetter coasts, drier mid-continent areas, and further sea level rise. Such changes could result in the gravest effects of climate change through human migration. Millions might be displaced by shoreline erosions, river and coastal flooding, or severe drought.

Migration related to climate change is likely to be predominantly from rural areas in developing countries to towns and cities. In the short term climate stress is likely to add incrementally to existing migration patterns rather than generating entirely new flows of people.

It has been argued that environmental degradation, loss of access to resources (e.g., water resources), and resulting human migration could become a source of political and even military conflict. Factors other than climate change may, however, be more important in affecting conflict. For example, Wilbanks *et al.* (2007) suggested that major environmentally influenced conflicts in Africa were more to do with the relative abundance of resources, e.g., oil and diamonds, than with resource scarcity. Scott *et al.* (2001) placed only low confidence in predictions of increased conflict due to climate change.

A 2013 study found that significant climatic changes were associated with a higher risk of conflict worldwide, and predicted that "amplified rates of human conflict could represent a large and critical social impact of anthropogenic climate change in both low- and high-income countries." Similarly, a 2014 study found that higher temperatures were associated with a greater likelihood of violent crime, and predicted that global warming would cause millions of such crimes in the United States alone during the 21st century.

Military planners are concerned that global warming is a "threat multiplier". "Whether it is poverty, food and water scarcity, diseases, economic instability, or threat of natural disasters, the broad range of changing climatic conditions may be far reaching. These challenges may threaten stability in much of the world".

Aggregate Impacts

Aggregating impacts adds up the total impact of climate change across sectors and/ or regions. Examples of aggregate measures include economic cost (e.g., changes in gross domestic product (GDP) and the social cost of carbon), changes in ecosystems (e.g., changes over land area from one type of vegetation to another), human health impacts, and the number of people affected by climate change. Aggregate measures such as economic cost require researchers to make value judgements over the importance of impacts occurring in different regions and at different times.

Observed Impacts

Global losses reveal rapidly rising costs due to extreme weather-related events since the 1970s. Socio-economic factors have contributed to the observed trend of global losses, e.g., population growth, increased wealth. Part of the growth is also related to regional climatic factors, e.g., changes in precipitation and flooding events. It is difficult to quantify the relative impact of socio-economic factors and climate change on the observed trend. The trend does, however, suggest increasing vulnerability of social systems to climate change.

Projected Impacts

The total economic impacts from climate change are highly uncertain. With medium confidence, Smith *et al.* (2001) concluded that world GDP would change by plus or minus a few percent for a small increase in global mean temperature (up to around 2 °C relative to the 1990 temperature level). Most studies assessed by Smith *et al.* (2001) projected losses in world GDP for a medium increase in global mean temperature (above 2-3 °C relative to the 1990 temperature level), with increasing losses for greater temperature increases. This assessment is consistent with the findings of more recent studies, as reviewed by Hitz and Smith (2004).

Economic impacts are expected to vary regionally. For a medium increase in global mean temperature (2-3 °C of warming, relative to the average temperature between 1990–2000), market sectors in low-latitude and less-developed areas might experience net costs due to climate change. On the other hand, market sectors in high-latitude and developed regions might experience net benefits for this level of warming. A global mean temperature increase above about 2-3 °C (relative to 1990-2000) would very likely result in market sectors across all regions experiencing either declines in net benefits or rises in net costs.

Aggregate impacts have also been quantified in non-economic terms. For example, climate change over the 21st century is likely to adversely affect hundreds of millions of people through increased coastal flooding, reductions in water supplies, increased malnutrition and increased health impacts.

Biological Systems

Observed Impacts on Biological Systems

A vast array of physical and biological systems across the Earth are being affected by human-induced global warming.

With very high confidence, Rosenzweig *et al.* (2007) concluded that recent warming had strongly affected natural biological systems. Hundreds of studies have documented responses of ecosystems, plants, and animals to the climate changes that have already occurred. For example, in the Northern Hemisphere, species are almost uniformly moving their ranges northward and up in elevation in search of cooler temperatures. Humans are very likely causing changes in regional temperatures to which plants and animals are responding.

Projected Impacts on Biological Systems

By the year 2100, ecosystems will be exposed to atmospheric CO_2 levels substantially higher than in the past 650,000 years, and global temperatures at least among the highest of those experienced in the past 740,000 years. Significant disruptions of ecosystems are projected to increase with future climate change. Examples of disruptions include disturbances such as fire, drought, pest infestation, invasion of species, storms, and coral bleaching events. The stresses caused by climate change, added to other stresses on ecological systems (e.g., land conversion, land degradation, harvesting, and pollution), threaten substantial damage to or complete loss of some unique ecosystems, and extinction of some critically endangered species.

Climate change has been estimated to be a major driver of biodiversity loss in cool conifer forests, savannas, mediterranean-climate systems, tropical forests, in the Arctic tundra, and in coral reefs. In other ecosystems, land-use change may be a stronger driver of biodiversity loss at least in the near-term. Beyond the year 2050, climate change may be the major driver for biodiversity loss globally.

A literature assessment by Fischlin *et al.* (2007) included a quantitative estimate of the number of species at increased risk of extinction due to climate change. With medium confidence, it was projected that approximately 20 to 30% of plant and animal species assessed so far (in an unbiased sample) would likely be at increasingly high risk of extinction should global mean temperatures exceed a warming of 2 to 3 °C above pre-industrial temperature levels. The uncertainties in this estimate, however, are large: for a rise of about 2 °C the percentage may be as low as 10%, or for about 3 °C, as high as 40%, and depending on biota (all living organisms of an area, the flora and fauna considered as a unit) the range is between 1% and 80%. As global average temperature exceeds 4 °C above pre-industrial levels, model projections suggested that there could be significant extinctions (40-70% of species that were assessed) around the globe.

Assessing whether future changes in ecosystems will be beneficial or detrimental is largely based on how ecosystems are valued by human society. For increases in global average temperature exceeding 1.5 to 2.5 °C (relative to global temperatures over the years 1980-1999) and in concomitant atmospheric CO2 concentrations, projected changes in ecosystems will have predominantly negative consequences for biodiversity and ecosystems goods and services, e.g., water and food supply.

Abrupt or Irreversible Changes

Physical, ecological and social systems may respond in an abrupt, non-linear or irregular way to climate change. This is as opposed to a smooth or regular response. A quantitative entity behaves "irregularly" when its dynamics are discontinuous (i.e., not smooth), nondifferentiable, unbounded, wildly varying, or otherwise ill-defined. Such behaviour is often termed "singular". Irregular behaviour in Earth systems may give rise to certain thresholds, which, when crossed, may lead to a large change in the system.

Some singularities could potentially lead to severe impacts at regional or global scales. Examples of "large-scale" singularities are discussed on abrupt climate change, climate change feedback and runaway climate change. It is possible that human-induced climate change could trigger large-scale singularities, but the probabilities of triggering such events are, for the most part, poorly understood.

With low to medium confidence, Smith *et al.* (2001) concluded that a rapid warming of more than 3 °C above 1990 levels would exceed thresholds that would lead to large-scale discontinuities in the climate system. Since the assessment by Smith *et al.* (2001), improved scientific understanding provides more guidance for two large-scale singularities: the role of carbon cycle feedbacks in future climate change and the melting of the Greenland and West Antarctic ice sheets.

Biogeochemical Cycles

Climate change may have an effect on the carbon cycle in an interactive "feedback" process. A feedback exists where an initial process triggers changes in a second process that in turn influences the initial process. A positive feedback intensifies the original process, and a negative feedback reduces it. Models suggest that the interaction of the climate system and the carbon cycle is one where the feedback effect is positive.

Using the A2 SRES emissions scenario, Schneider *et al.* (2007) found that this effect led to additional warming by the years 2090-2100 (relative to the 1990–2000) of 0.1–1.5 °C. This estimate was made with high confidence. The climate projections made in the IPCC Fourth Assessment Report summarized earlier of 1.1–6.4 °C account for this feedback effect. On the other hand, with medium confidence, Schneider *et al.* (2007) commented that additional releases of GHGs were possible from permafrost, peat lands, wetlands, and large stores of marine hydrates at high latitudes.

Greenland and West Antarctic Ice Sheets

With medium confidence, authors of AR4 concluded that with a global average temperature increase of 1–4 °C (relative to temperatures over the years 1990–2000), at least a partial deglaciation of the Greenland ice sheet, and possibly the West Antarctic ice sheets would occur. The estimated timescale for partial deglaciation was centuries to millennia, and would contribute 4 to 6 metres (13 to 20 ft) or more to sea level rise over this period.

Atlantic Meridional Overturning Circulation

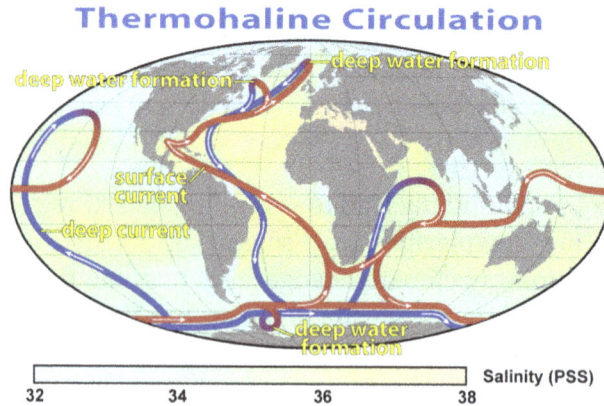

This map shows the general location and direction of the warm surface (red) and cold deep water (blue) currents of the thermohaline circulation. Salinity is represented by color in units of the Practical Salinity Scale. Low values (blue) are less saline, while high values (orange) are more saline.

The Atlantic Meridional Overturning Circulation (AMOC) is an important component of the Earth's climate system, characterized by a northward flow of warm, salty water in the upper layers of the Atlantic and a southward flow of colder water in the deep Atlantic. The AMOC is equivalently known as the thermohaline circulation (THC). Potential impacts associated with MOC changes include reduced warming or (in the case of abrupt change) absolute cooling of northern high-latitude areas near Greenland and north-western Europe, an increased warming of Southern Hemisphere high-latitudes, tropical drying, as well as changes to marine ecosystems, terrestrial vegetation, oceanic CO_2 uptake, oceanic oxygen concentrations, and shifts in fisheries. According to an assessment by the US Climate Change Science Program (CCSP, 2008b), it is very likely (greater than 90% probability, based on expert judgement) that the strength of the AMOC will decrease over the course of the 21st century. Warming is still expected to occur over most of the European region downstream of the North Atlantic Current in response to increasing GHGs, as well as over North America. Although it is very unlikely (less than 10% probability, based on expert judgement) that the AMOC will collapse in the 21st century, the potential consequences of such a collapse could be severe.

Irreversibilities

Commitment to Radiative Forcing

Emissions of GHGs are a potentially irreversible commitment to sustained radiative forcing in the future. The contribution of a GHG to radiative forcing depends on the gas's ability to trap infrared (heat) radiation, the concentration of the gas in the atmosphere, and the length of time the gas resides in the atmosphere.

CO_2 is the most important anthropogenic GHG. While more than half of the CO_2 emitted is currently removed from the atmosphere within a century, some fraction (about

20%) of emitted CO_2 remains in the atmosphere for many thousands of years. Consequently, CO_2 emitted today is potentially an irreversible commitment to sustained radiative forcing over thousands of years.

This commitment may not be truly irreversible should techniques be developed to remove CO_2 or other GHGs directly from the atmosphere, or to block sunlight to induce cooling. Techniques of this sort are referred to as geoengineering. Little is known about the effectiveness, costs or potential side-effects of geoengineering options. Some geoengineering options, such as blocking sunlight, would not prevent further ocean acidification.

Irreversible Impacts

Human-induced climate change may lead to irreversible impacts on physical, biological, and social systems. There are a number of examples of climate change impacts that may be irreversible, at least over the timescale of many human generations. These include the large-scale singularities described above – changes in carbon cycle feedbacks, the melting of the Greenland and West Antarctic ice sheets, and changes to the AMOC. In biological systems, the extinction of species would be an irreversible impact. In social systems, unique cultures may be lost due to climate change. For example, humans living on atoll islands face risks due to sea-level rise, sea-surface warming, and increased frequency and intensity of extreme weather events.

Benefits of Global Warming

In some locations and industries global warming may increase productivity, though the IPCC cautions that "Estimates agree on the size of the impact (small relative to economic growth), and 17 of the 20 impact estimates shown are negative. Losses accelerate with greater warming, and estimates diverge." The identified benefits are listed below.

CO2 Fertilisation Effect

CO2 is one of the substances which plants require to grow. Increasing its amount in the air contributes to:

- Improved agriculture in some high latitude regions
- Increased growing season in Greenland
- Increased productivity of sour orange trees
- Increased vegetation activity in high northern latitudes
- Increased plankton biomass in the North Pacific Subtropical Gyre
- Recent increase in forest growth
- Increased Arctic tundra plant reproduction

Human Health

- Winter deaths might decline as temperatures warm. However, this is disputed for at least some regions. For example states "Although excess winter deaths evidently do exist, winter cold severity no longer predicts the numbers affected. We conclude that no evidence exists that excess winter deaths in England and Wales will fall if winters warm with climate change".

Ice-free Northwest Passage

- Ships will travel on a shorter route between the Pacific and Atlantic oceans.

Animal Population Changes

Some animals will benefit from the warming:

- Increase in chinstrap and gentoo penguins.

- Bigger marmots.

Scientific Opinion

The Intergovernmental Panel on Climate Change (IPCC) has published several major assessments on the effects of global warming. Its most recent comprehensive impact assessment was published in 2014. Publications describing the effects of climate change have also been produced by the following organizations:

- American Association for the Advancement of Science (AAAS)

- A report by the Netherlands Environmental Assessment Agency, the Royal Netherlands Meteorological Institute, and Wageningen University and Research Centre

- UK AVOID research programme

- A report by the UK Royal Society and US National Academy of Sciences

- University of New South Wales Climate Change Research Centre

- US National Research Council

A report by Molina *et al.* states:

The overwhelming evidence of human-caused climate change documents both current impacts with significant costs and extraordinary future risks to society and natural systems

NASA Data and Tools

NASA has released public data and tools to predict how temperature and rainfall patterns worldwide may change through to the year 2100 caused by increasing carbon dioxide in Earth's atmosphere. The dataset shows projected changes worldwide on a regional level simulated by 21 climate models. The data can be viewed on a daily times-cale for individual cities and towns and may be used to conduct climate risk assessments to predict the local and global effects of weather dangers, for example droughts, floods, heat waves and declines in agriculture productivity, and help plan responses to global warming effects.

Climate Change

Climate change is a change in the statistical distribution of weather patterns when that change lasts for an extended period of time (i.e., decades to millions of years). Climate change may refer to a change in average weather conditions, or in the time variation of weather around longer-term average conditions (i.e., more or fewer extreme weather events). Climate change is caused by factors such as biotic processes, variations in solar radiation received by Earth, plate tectonics, and volcanic eruptions. Certain human activities have been identified as primary causes of ongoing climate change, often referred to as *global warming*.

Scientists actively work to understand past and future climate by using observations and theoretical models. A climate record extending deep into the Earth's past has been assembled, and continues to be built up, based on geological evidence from borehole temperature profiles, cores removed from deep accumulations of ice, floral and faunal records, glacial and periglacial processes, stable-isotope and other analyses of sediment layers, and records of past sea levels. More recent data are provided by the instrumental record. General circulation models, based on the physical sciences, are often used in theoretical approaches to match past climate data, make future projections, and link causes and effects in climate change.

Terminology

The most general definition of *climate change* is a change in the statistical properties (principally its mean and spread) of the climate system when considered over long periods of time, regardless of cause. Accordingly, fluctuations over periods shorter than a few decades, such as El Niño, do not represent climate change.

The term "climate change" is often used to refer specifically to *anthropogenic* climate change (also known as global warming). Anthropogenic climate change is caused by human activity, as opposed to changes in climate that may have resulted as part of Earth's natural processes. In this sense, especially in the context of environmental policy, the term *climate change* has become synonymous with *anthropogenic global warming*. Within scientific journals, *global warming* refers to surface temperature increases while *climate change* includes global warming and everything else that increasing greenhouse gas levels affect.

A related term is "climatic change". In 1966, the World Meteorological Organization (WMO) proposed the term "climatic change" to encompass all forms of climatic variability on timescales longer than 10 years, regardless of cause. Change was a given and climatic was used as an adjective to describe this kind of change (as opposed to political or economic change). When it was realized that human activities had a potential to drastically alter the climate, the term climate change replaced climatic change as the dominant term to reflect an anthropogenic cause. Climate change was incorporated in the title of the Intergovernmental Panel on Climate Change (IPCC) and the UN Framework Convention on Climate Change (UNFCCC). Climate change, used as a noun, became an issue rather than the technical description of changing weather.

Causes

On the broadest scale, the rate at which energy is received from the Sun and the rate at which it is lost to space determine the equilibrium temperature and climate of Earth. This energy is distributed around the globe by winds, ocean currents, and other mechanisms to affect the climates of different regions.

Factors that can shape climate are called climate forcings or "forcing mechanisms". These include processes such as variations in solar radiation, variations in the Earth's orbit, variations in the albedo or reflectivity of the continents, atmosphere, and oceans, mountain-building and continental drift and changes in greenhouse gas concentrations. There are a variety of climate change feedbacks that can either amplify or diminish the initial forcing. Some parts of the climate system, such as the oceans and ice caps, respond more slowly in reaction to climate forcings, while others respond more quickly. There are also key threshold factors which when exceeded can produce rapid change.

Forcing mechanisms can be either "internal" or "external". Internal forcing mechanisms are natural processes within the climate system itself (e.g., the thermohaline circulation). External forcing mechanisms can be either natural (e.g., changes in solar output, the earth's orbit, volcano eruptions) or anthropogenic (e.g. increased emissions of greenhouse gases and dust).

Whether the initial forcing mechanism is internal or external, the response of the climate system might be fast (e.g., a sudden cooling due to airborne volcanic ash reflecting sunlight), slow (e.g. thermal expansion of warming ocean water), or a combination (e.g., sudden loss of albedo in the Arctic Ocean as sea ice melts, followed by more gradual thermal expansion of the water). Therefore, the climate system can respond abruptly, but the full response to forcing mechanisms might not be fully developed for centuries or even longer.

Internal Forcing Mechanisms

Scientists generally define the five components of earth's climate system to include atmosphere, hydrosphere, cryosphere, lithosphere (restricted to the surface soils, rocks, and sediments), and biosphere. Natural changes in the climate system ("internal forcings") result in internal "climate variability". Examples include the type and distribution of species, and changes in ocean-atmosphere circulations.

Ocean-atmosphere Variability

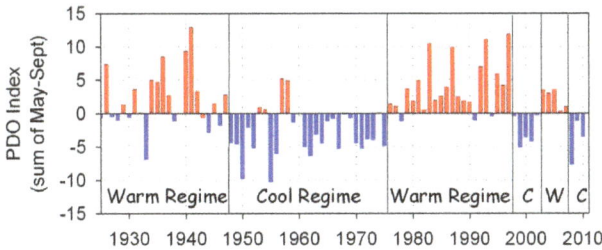

Pacific Decadal Oscillation 1925 to 2010

The ocean and atmosphere can work together to spontaneously generate internal climate variability that can persist for years to decades at a time. Examples of this type of variability include the El Niño-Southern Oscillation, the Pacific decadal oscillation, and the Atlantic Multidecadal Oscillation. These variations can affect global average surface temperature by redistributing heat between the deep ocean and the atmopshere and/or by altering the cloud/water vapor/sea ice distribution which can affect the total energy budget of the earth.

The oceanic aspects of these circulations can generate variability on centennial timescales due to the ocean having hundreds of times more mass than in the atmosphere, and thus very high thermal inertia. For example, alterations to ocean processes such as thermohaline circulation play a key role in redistributing heat in the world's oceans. Due to the long timescales of this circulation, ocean temperature at depth is still adjusting to effects of the Little Ice Age which occurred between the 1600 and 1800s.

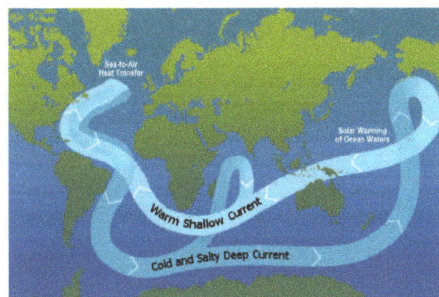

A schematic of modern thermohaline circulation. Tens of millions of years ago, continental-plate movement formed a land-free gap around Antarctica, allowing the formation of the ACC, which keeps warm waters away from Antarctica.

Life

Life affects climate through its role in the carbon and water cycles and through such mechanisms as albedo, evapotranspiration, cloud formation, and weathering. Examples of how life may have affected past climate include:

- glaciation 2.3 billion years ago triggered by the evolution of oxygenic photosynthesis, which depleted the atmosphere of the greenhouse gas carbon dioxide and introduced free oxygen.

- another glaciation 300 million years ago ushered in by long-term burial of decomposition-resistant detritus of vascular land-plants (creating a carbon sink and forming coal)

- termination of the Paleocene-Eocene Thermal Maximum 55 million years ago by flourishing marine phytoplankton

- reversal of global warming 49 million years ago by 800,000 years of arctic azolla blooms

- global cooling over the past 40 million years driven by the expansion of grass-grazer ecosystems

External Forcing Mechanisms

Milankovitch cycles from 800,000 years ago in the past to 800,000 years in the future.

Variations in CO_2, temperature and dust from the Vostok ice core over the last 450,000 years.

Orbital Variations

Slight variations in Earth's orbit lead to changes in the seasonal distribution of sunlight reaching the Earth's surface and how it is distributed across the globe. There is very little change to the area-averaged annually averaged sunshine; but there can be strong changes in the geographical and seasonal distribution. The three types of orbital variations are variations in Earth's eccentricity, changes in the tilt angle of Earth's axis of rotation, and precession of Earth's axis. Combined together, these produce Milankovitch cycles which have a large impact on climate and are notable for their correlation to glacial and interglacial periods, their correlation with the advance and retreat of the Sahara, and for their appearance in the stratigraphic record.

The IPCC notes that Milankovitch cycles drove the ice age cycles, CO_2 followed temperature change "with a lag of some hundreds of years", and that as a feedback amplified temperature change. The depths of the ocean have a lag time in changing temperature (thermal inertia on such scale). Upon seawater temperature change, the solubility of CO_2 in the oceans changed, as well as other factors impacting air-sea CO_2 exchange.

Solar Output

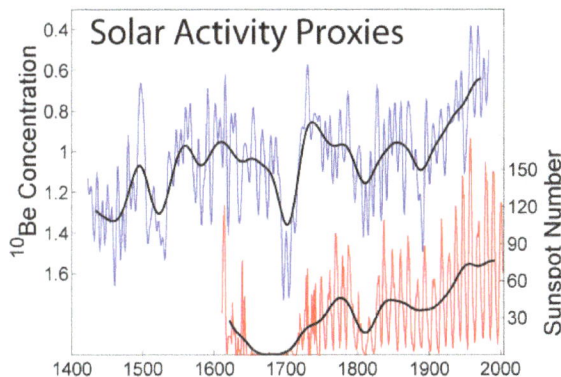

Variations in solar activity during the last several centuries based on observations of sunspots and beryllium isotopes. The period of extraordinarily few sunspots in the late 17th century was the Maunder minimum.

The Sun is the predominant source of energy input to the Earth. Other sources include geothermal energy from the Earth's core, and heat from the decay of radioactive compounds. Both long- and short-term variations in solar intensity are known to affect global climate.

Three to four billion years ago, the Sun emitted only 70% as much power as it does today. If the atmospheric composition had been the same as today, liquid water should not have existed on Earth. However, there is evidence for the presence of water on the early Earth, in the Hadean and Archean eons, leading to what is known as the faint young Sun paradox. Hypothesized solutions to this paradox include a vastly different atmosphere, with much higher concentrations of greenhouse gases than currently exist. Over the following approximately 4 billion years, the energy output of the Sun increased and atmospheric composition changed. The Great Oxygenation Event – oxygenation of the atmosphere around 2.4 billion years ago – was the most notable alteration. Over the next five billion years, the Sun's ultimate death as it becomes a red giant and then a white dwarf will have large effects on climate, with the red giant phase possibly ending any life on Earth that survives until that time.

Solar output also varies on shorter time scales, including the 11-year solar cycle and longer-term modulations. Solar intensity variations possibly as a result of the Wolf, Spörer and Maunder Minimum are considered to have been influential in triggering the Little Ice Age, and some of the warming observed from 1900 to 1950. The cyclical nature of the Sun's energy output is not yet fully understood; it differs from the very slow change that is happening within the Sun as it ages and evolves. Research indicates that solar variability has had effects including the Maunder minimum from 1645 to 1715 A.D., part of the Little Ice Age from 1550 to 1850 A.D. that was marked by relative cooling and greater glacier extent than the centuries before and afterward. Some studies point toward solar radiation increases from cyclical sunspot activity affecting global warming, and climate may be influenced by the sum of all effects (solar variation, anthropogenic radiative forcings, etc.).

Interestingly, a 2010 study *suggests*, "that the effects of solar variability on temperature throughout the atmosphere may be contrary to current expectations."

In an Aug 2011 Press Release, CERN announced the publication in the Nature journal the initial results from its CLOUD experiment. The results indicate that ionisation from cosmic rays significantly enhances aerosol formation in the presence of sulfuric acid and water, but in the lower atmosphere where ammonia is also required, this is insufficient to account for aerosol formation and additional trace vapours must be involved. The next step is to find more about these trace vapours, including whether they are of natural or human origin.

Volcanism

The eruptions considered to be large enough to affect the Earth's climate on a scale of more than 1 year are the ones that inject over 100,000 tons of SO_2 into the stratosphere. This is due to the optical properties of SO_2 and sulfate aerosols, which strongly absorb or scatter solar radiation, creating a global layer of sulfuric acid haze. On average, such eruptions occur several times per century, and cause cooling (by partially blocking the transmission of solar radiation to the Earth's surface) for a period of a few years.

The eruption of Mount Pinatubo in 1991, the second largest terrestrial eruption of the 20th century, affected the climate substantially, subsequently global temperatures decreased by about 0.5 °C (0.9 °F) for up to three years. Thus, the cooling over large parts of the Earth reduced surface temperatures in 1991–93, the equivalent to a reduction in net radiation of 4 watts per square meter. The Mount Tambora eruption in 1815 caused the Year Without a Summer. Much larger eruptions, known as large igneous provinces, occur only a few times every fifty – one hundred million years – through flood basalt, and caused in Earth past global warming and mass extinctions.

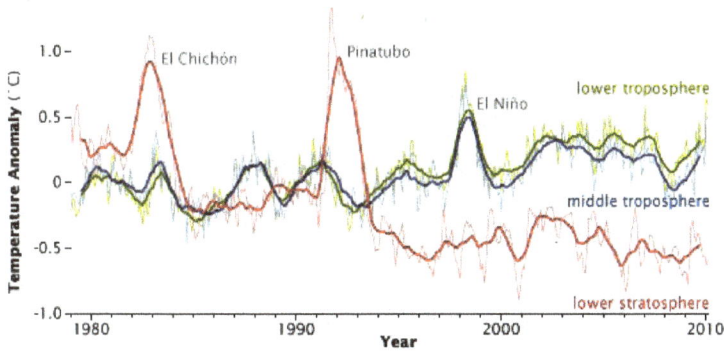

In atmospheric temperature from 1979 to 2010, determined by MSU NASA satellites, effects appear from aerosols released by major volcanic eruptions (El Chichón and Pinatubo). El Niño is a separate event, from ocean variability.

Small eruptions, with injections of less than 0.1 Mt of sulfur dioxide into the stratosphere, impact the atmosphere only subtly, as temperature changes are comparable with natural variability. However, because smaller eruptions occur at a much higher frequency, they too have a significant impact on Earth's atmosphere.

Seismic monitoring maps current and future trends in volcanic activities, and tries to develop early warning systems. In climate modelling the aim is to study the physical mechanisms and feedbacks of volcanic forcing.

Volcanoes are also part of the extended carbon cycle. Over very long (geological) time periods, they release carbon dioxide from the Earth's crust and mantle, counteracting the uptake by sedimentary rocks and other geological carbon dioxide sinks. The US Geological Survey estimates are that volcanic emissions are at a much lower level than the effects of current human activities, which generate 100–300 times the amount of carbon dioxide emitted by volcanoes. A review of published studies indicates that annual volcanic emissions of carbon dioxide, including amounts released from mid-ocean ridges, volcanic arcs, and hot spot volcanoes, are only the equivalent of 3 to 5 days of human-caused output. The annual amount put out by human activities may be greater than the amount released by supererruptions, the most recent of which was the Toba eruption in Indonesia 74,000 years ago.

Although volcanoes are technically part of the lithosphere, which itself is part of the climate system, the IPCC explicitly defines volcanism as an external forcing agent.

Plate Tectonics

Over the course of millions of years, the motion of tectonic plates reconfigures global land and ocean areas and generates topography. This can affect both global and local patterns of climate and atmosphere-ocean circulation.

The position of the continents determines the geometry of the oceans and therefore influences patterns of ocean circulation. The locations of the seas are important in controlling the transfer of heat and moisture across the globe, and therefore, in determining global climate. A recent example of tectonic control on ocean circulation is the formation of the Isthmus of Panama about 5 million years ago, which shut off direct mixing between the Atlantic and Pacific Oceans. This strongly affected the ocean dynamics of what is now the Gulf Stream and may have led to Northern Hemisphere ice cover. During the Carboniferous period, about 300 to 360 million years ago, plate tectonics may have triggered large-scale storage of carbon and increased glaciation. Geologic evidence points to a "megamonsoonal" circulation pattern during the time of the supercontinent Pangaea, and climate modeling suggests that the existence of the supercontinent was conducive to the establishment of monsoons.

The size of continents is also important. Because of the stabilizing effect of the oceans on temperature, yearly temperature variations are generally lower in coastal areas than they are inland. A larger supercontinent will therefore have more area in which climate is strongly seasonal than will several smaller continents or islands.

Human Influences

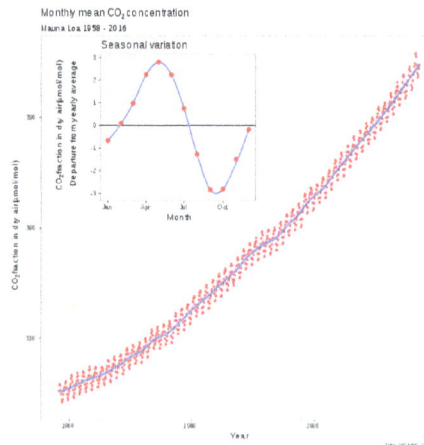

Increase in atmospheric CO_2 levels

In the context of climate variation, anthropogenic factors are human activities which affect the climate. The scientific consensus on climate change is "that climate is changing and that these changes are in large part caused by human activities," and it "is largely irreversible."

"Science has made enormous inroads in understanding climate change and its causes, and is beginning to help develop a strong understanding of current and potential impacts that will affect people today and in coming decades. This understanding is crucial because it allows decision makers to place climate change in the context of other large challenges facing the nation and the world. There are still some uncertainties, and there always will be in understanding a complex system like Earth's climate. Nevertheless, there is a strong, credible body of evidence, based on multiple lines of research, documenting that climate is changing and that these changes are in large part caused by human activities. While much remains to be learned, the core phenomenon, scientific questions, and hypotheses have been examined thoroughly and have stood firm in the face of serious scientific debate and careful evaluation of alternative explanations."

— United States National Research Council, Advancing the Science of Climate Change

Of most concern in these anthropogenic factors is the increase in CO_2 levels. This is due to emissions from fossil fuel combustion, followed by aerosols (particulate matter in the atmosphere), and the CO_2 released by cement manufacture. Other factors, including land use, ozone depletion, animal husbandry (ruminant animals such as cattle produce methane, as do termites as well), and deforestation, are also of concern in the roles they play – both separately and in conjunction with other factors – in affecting climate, microclimate, and measures of climate variables.

Physical Evidence

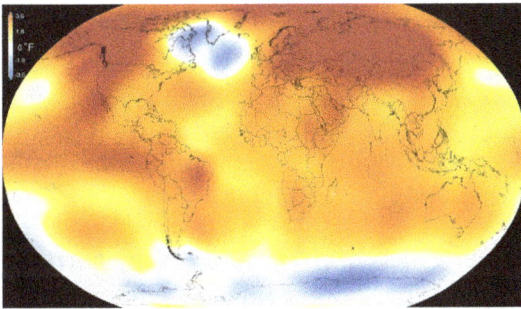

Global temperature anomalies for 2015 compared to the 1951-1980 baseline. 2015 was the warmest year in the NASA/NOAA temperature record with starts in 1880. It has since been superseded by 2016.

Evidence for climatic change is taken from a variety of sources that can be used to reconstruct past climates. Reasonably complete global records of surface temperature are available beginning from the mid-late 19th century. For earlier periods, most of the evidence is indirect—climatic changes are inferred from changes in proxies, indicators that reflect climate, such as vegetation, ice cores, dendrochronology, sea level change, and glacial geology.

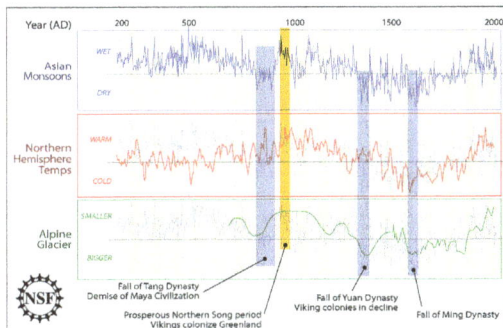

Comparisons between Asian Monsoons from 200 AD to 2000 AD (staying in the background on other plots), Northern Hemisphere temperature, Alpine glacier extent (vertically inverted as marked), and human history as noted by the U.S. NSF.

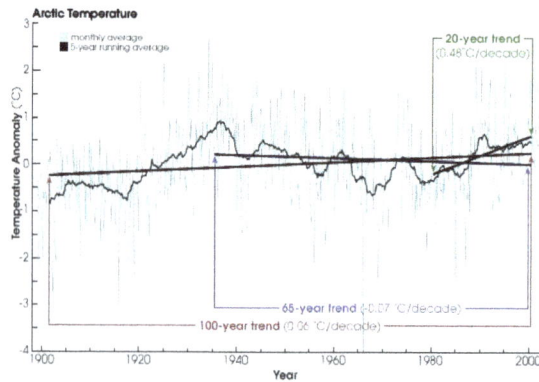

Arctic temperature anomalies over a 100-year period as estimated by NASA. Typical high monthly variance can be seen, while longer-term averages highlight trends.

Temperature Measurements and Proxies

The instrumental temperature record from surface stations was supplemented by radiosonde balloons, extensive atmospheric monitoring by the mid-20th century, and, from the 1970s on, with global satellite data as well. The $^{18}O/^{16}O$ ratio in calcite and ice core samples used to deduce ocean temperature in the distant past is an example of a temperature proxy method, as are other climate metrics noted in subsequent categories.

Historical and Archaeological Evidence

Climate change in the recent past may be detected by corresponding changes in settlement and agricultural patterns. Archaeological evidence, oral history and historical documents can offer insights into past changes in the climate. Climate change effects have been linked to the collapse of various civilizations.

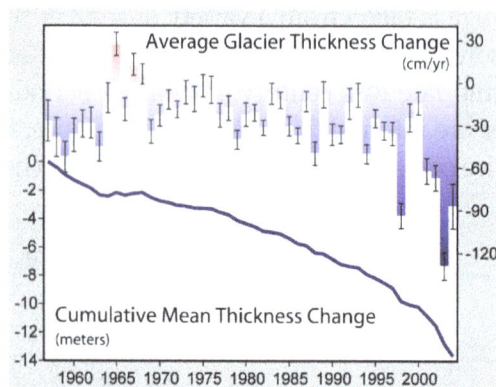

Decline in thickness of glaciers worldwide over the past half-century

Glaciers

Glaciers are considered among the most sensitive indicators of climate change. Their size is determined by a mass balance between snow input and melt output. As temperatures warm, glaciers retreat unless snow precipitation increases to make up for the additional melt; the converse is also true.

Glaciers grow and shrink due both to natural variability and external forcings. Variability in temperature, precipitation, and englacial and subglacial hydrology can strongly determine the evolution of a glacier in a particular season. Therefore, one must average over a decadal or longer time-scale and/or over many individual glaciers to smooth out the local short-term variability and obtain a glacier history that is related to climate.

A world glacier inventory has been compiled since the 1970s, initially based mainly on aerial photographs and maps but now relying more on satellites. This compilation tracks more than 100,000 glaciers covering a total area of approximately 240,000 km², and preliminary estimates indicate that the remaining ice cover is around 445,000 km². The World Glacier Monitoring Service collects data annually on glacier retreat and glacier mass balance. From this data, glaciers worldwide have been found to be shrinking significantly, with strong glacier retreats in the 1940s, stable or growing conditions during the 1920s and 1970s, and again retreating from the mid-1980s to present.

The most significant climate processes since the middle to late Pliocene (approximately 3 million years ago) are the glacial and interglacial cycles. The present interglacial period (the Holocene) has lasted about 11,700 years. Shaped by orbital variations, responses such as the rise and fall of continental ice sheets and significant sea-level changes helped create the climate. Other changes, including Heinrich events, Dansgaard–Oeschger events and the Younger Dryas, however, illustrate how glacial variations may also influence climate without the orbital forcing.

Glaciers leave behind moraines that contain a wealth of material—including organic matter, quartz, and potassium that may be dated—recording the periods in which a glacier advanced and retreated. Similarly, by tephrochronological techniques, the lack of glacier cover can be identified by the presence of soil or volcanic tephra horizons whose date of deposit may also be ascertained.

Arctic Sea Ice Loss

The decline in Arctic sea ice, both in extent and thickness, over the last several decades is further evidence for rapid climate change. Sea ice is frozen seawater that floats on the ocean surface. It covers millions of square miles in the polar regions, varying with the seasons. In the Arctic, some sea ice remains year after year, whereas almost all Southern Ocean or Antarctic sea ice melts away and reforms annually. Satellite observations show that Arctic sea ice is now declining at a rate of 13.3 percent per decade, relative to the 1981 to 2010 average.

Vegetation

A change in the type, distribution and coverage of vegetation may occur given a change in the climate. Some changes in climate may result in increased precipitation and warmth, resulting in improved plant growth and the subsequent sequestration of airborne CO_2. A gradual increase in warmth in a region will lead to earlier flowering and fruiting times, driving a change in the timing of life cycles of dependent organisms. Conversely, cold will cause plant bio-cycles to lag. Larger, faster or more radical changes, however, may result in vegetation stress, rapid plant loss and desertification in certain circumstances. An example of this occurred during the Carboniferous Rainforest Collapse (CRC), an extinction event 300 million years ago. At this time vast rainforests covered the equatorial region of Europe and America. Climate change devastated these tropical rainforests, abruptly fragmenting the habitat into isolated 'islands' and causing the extinction of many plant and animal species.

Forest Genetic Resources

Even though this is a field with many uncertainties, it is expected that over the next 50 years climate changes will have an effect on the diversity of forest genetic resources and thereby on the distribution of forest tree species and the composition of forests. Diversity of forest genetic resources enables the potential for a species (or a population) to adapt to climatic changes and related future challenges such as temperature changes, drought, pests, diseases and forest fire. However, species are not naturally capable to adapt in the pace of which the climate is changing and the increasing temperatures will most likely facilitate the spread of pests and diseases, creating an additional threat to forest trees and their populations. To inhibit these problems human interventions, such as transfer of forest reproductive material, may be needed.

Pollen Analysis

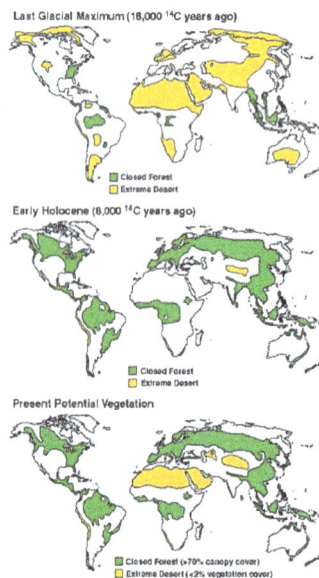

Top: Arid ice age climate
Middle: Atlantic Period, warm and wet
Bottom: Potential vegetation in climate now if not for human effects like agriculture.

Palynology is the study of contemporary and fossil palynomorphs, including pollen. Palynology is used to infer the geographical distribution of plant species, which vary under different climate conditions. Different groups of plants have pollen with distinctive shapes and surface textures, and since the outer surface of pollen is composed of a very resilient material, they resist decay. Changes in the type of pollen found in different layers of sediment in lakes, bogs, or river deltas indicate changes in plant communities. These changes are often a sign of a changing climate. As an example, palynological studies have been used to track changing vegetation patterns throughout the Quaternary glaciations and especially since the last glacial maximum.

Cloud Cover and Precipitation

Past precipitation can be estimated in the modern era with the global network of precipitation gauges. Surface coverage over oceans and remote areas is relatively sparse, but, reducing reliance on interpolation, satellite clouds and precipitation data has been available since the 1970s. Quantification of climatological variation of precipitation in prior centuries and epochs is less complete but approximated using proxies such as marine sediments, ice cores, cave stalagmites, and tree rings. In July 2016 scientists published evidence of increased cloud cover over polar regions, as predicted by climate models.

Climatological temperatures substantially affect cloud cover and precipitation. For instance, during the Last Glacial Maximum of 18,000 years ago, thermal-driven evaporation from the oceans onto continental landmasses was low, causing large areas of extreme desert, including polar deserts (cold but with low rates of cloud cover and precipitation). In contrast, the world's climate was cloudier and wetter than today near the start of the warm Atlantic Period of 8000 years ago.

Estimated global land precipitation increased by approximately 2% over the course of the 20th century, though the calculated trend varies if different time endpoints are chosen, complicated by ENSO and other oscillations, including greater global land cloud cover precipitation in the 1950s and 1970s than the later 1980s and 1990s despite the positive trend over the century overall. Similar slight overall increase in global river runoff and in average soil moisture has been perceived.

Dendroclimatology

Dendroclimatology is the analysis of tree ring growth patterns to determine past climate variations. Wide and thick rings indicate a fertile, well-watered growing period, whilst thin, narrow rings indicate a time of lower rainfall and less-than-ideal growing conditions.

Ice Cores

Analysis of ice in a core drilled from an ice sheet such as the Antarctic ice sheet, can be used to show a link between temperature and global sea level variations. The air trapped in bubbles in the ice can also reveal the CO_2 variations of the atmosphere from the distant past, well before modern environmental influences. The study of these ice cores has been a significant indicator of the changes in CO_2 over many millennia, and continues to provide valuable information about the differences between ancient and modern atmospheric conditions.

Animals

Remains of beetles are common in freshwater and land sediments. Different species of beetles tend to be found under different climatic conditions. Given the extensive lineage of beetles whose genetic makeup has not altered significantly over the millennia, knowledge of the present climatic range of the different species, and the age of the sediments in which remains are found, past climatic conditions may be inferred.

Similarly, the historical abundance of various fish species has been found to have a substantial relationship with observed climatic conditions. Changes in the primary productivity of autotrophs in the oceans can affect marine food webs.

Sea Level Change

Global sea level change for much of the last century has generally been estimated using tide gauge measurements collated over long periods of time to give a long-term average. More recently, altimeter measurements in combination with accurately determined satellite orbits have provided an improved measurement of global sea level change. To measure sea levels prior to instrumental measurements, scientists have dated coral reefs that grow near the surface of the ocean, coastal sediments, marine terraces, ooids in limestones, and nearshore archaeological remains. The predominant dating methods used are uranium series and radiocarbon, with cosmogenic radionuclides being sometimes used to date terraces that have experienced relative sea level fall. In the early Pliocene, global temperatures were 1–2 °C warmer than the present temperature, yet sea level was 15–25 meters higher than today.

Physical Impacts of Climate Change

In their usage, "climate change" refers to a change in the state of the climate that can be identified by changes in the mean and/or variability of its properties, and that persists for extended periods, typically decades or longer (IPCC, 2007d:30). The climate change referred to may be due to natural causes and/or the result of human activity.

Global Warming

Global surface temperatures have increased about 0.74 °C (plus or minus 0.18 °C) since the late-19th century, and the linear trend for the past 50 years of 0.13 °C (plus or minus 0.03 °C) per decade is nearly twice that for the past 100 years. The warming has not been globally uniform. Some areas have, in fact, cooled slightly over the last century. The recent warmth has been greatest over North America and Eurasia between 40 and 70°N. Lastly, seven of the eight warmest years on record have occurred since 2001 and the 10 warmest years have all occurred since 1995.

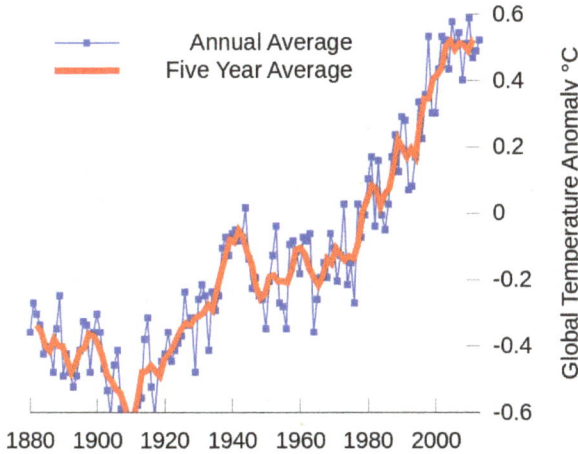

Global mean surface temperature difference from the average for 1880–2009.

Consistency of Evidence for Warming

Thousands of land and ocean temperature measurements are recorded each day around the globe. This includes measurements from climate reference stations, weather stations, ships, buoys and autonomous gliders in the oceans. These surface measurements are also supplemented with satellite measurements. These measurements are processed, examined for random and systematic errors, and then finally combined to produce a time series of global average temperature change. A number of agencies around the world have produced datasets of global-scale changes in surface temperature using different techniques to process the data and remove measurement errors that could lead to false interpretations of temperature trends. The warming trend that is apparent in all of the independent methods of calculating global temperature change is also confirmed by other independent observations, such as the melting of mountain glaciers on every continent, reductions in the extent of snow cover, earlier blooming of plants in spring, a shorter ice season on lakes and rivers, ocean heat content, reduced Arctic sea ice, and rising sea levels.

Global Average Temperature

Global average temperature is one of the most-cited indicators of global climate change, and shows an increase of approximately 1.4 °F since the early 20th Century. The global surface temperature is based on air temperature data over land and sea-surface temperatures observed from ships, buoys and satellites. There is a clear long-term global

warming trend, while each individual year does not always show a temperature increase relative to the previous year, and some years show greater changes than others.
These year-to-year fluctuations in temperature are due to natural processes, such as
the effects of El Niños, La Niñas, and the eruption of large volcanoes. Notably, the 20
warmest years have all occurred since 1981, and the 10 warmest have all occurred in
the past 12 years.

Diurnal Temperature

Annual anomalies of maximum and minimum temperatures and diurnal temperature
range (DTR) (°C) relative to the 1961 to 1990 mean, averaged for the 71 percent of
global land areas where data are available for 1950 to 2004. The smooth curves show
decadal variations.

There has been a general, but not global, tendency toward reduced diurnal temperature
range (DTR: the difference between daily high or maximum and daily low or minimum
temperatures) over about 70% of the global land mass since the middle of the 20th
century. However, for the period 1979–2005 the DTR shows no trend since the trend
in both maximum and minimum temperatures for the same period are virtually identi-
cal; both showing a strong warming signal. A variety of factors likely contribute to this
change in DTR, particularly on a regional and local basis, including changes in cloud
cover, atmospheric water vapor, land use and urban effects.

Indirect Indicators of Warming

Indirect indicators of warming such as borehole temperatures, snow cover, and glacier
recession data, are in substantial agreement with the more direct indicators of recent
warmth. Evidence such as changes in glacial mass balance (the amount of snow and
ice contained in a glacier) is useful since it not only provides qualitative support for
existing meteorological data, but glaciers often exist in places too remote to support
meteorological stations. The records of glacial advance and retreat often extend back
further than weather station records, and glaciers are usually at much higher altitudes
than weather stations, allowing scientists more insight into temperature changes higher in the atmosphere.

Effects on Weather

Increasing temperature is likely to lead to increasing precipitation but the effects on
storms are less clear. Extratropical storms partly depend on the temperature gradient,
which is predicted to weaken in the northern hemisphere as the polar region warms
more than the rest of the hemisphere. It is possible that the Polar and Ferrel cells in
one or both hemispheres will weaken and eventually disappear, which would cause the
Hadley cell to cover the whole planet. This would greatly decrease the temperature gradient between the arctic and the tropics, and cause the earth to flip to a hothouse state.

Precipitation

Projected change in annual average precipitation by the end of the 21st century, based on a medium emissions scenario (SRES A1B).

Historically (i.e., over the 20th century), subtropical land regions have been mostly semi-arid, while most subpolar regions have had an excess of precipitation over evaporation. Future global warming is expected to be accompanied by a reduction in rainfall in the subtropics and an increase in precipitation in subpolar latitudes and some equatorial regions. In other words, regions which are dry at present will generally become even drier, while regions that are currently wet will generally become even wetter. This projection does not apply to every locale, and in some cases can be modified by local conditions. Drying is projected to be strongest near the poleward margins of the subtropics (for example, South Africa, southern Australia, the Mediterranean, and the south-western U.S.), a pattern that can be described as a poleward expansion of these semi-arid zones.

This large-scale pattern of change is a robust feature present in nearly all of the simulations conducted by the world's climate modeling groups for the 4th Assessment of the Intergovernmental Panel on Climate Change (IPCC), and is also evident in observed 20th century precipitation trends.

Extreme Events

Fire

Fire is a major agent for conversion of biomass and soil organic matter to CO_2 (Denman *et al.*, 2007:527). There is a large potential for future alteration in the terrestrial carbon balance through altered fire regimes. With high confidence, Schneider *et al.* (2007) projected that:

- An increase in global mean temperature of about 0 to 2 °C by 2100 relative to the 1990–2000 period would result in increased fire frequency and intensity in many areas.

- An increase in the region of 2 °C or above would lead to increased frequency and intensity of fires.

Extreme Weather

IPCC (2007a:8) projected that in the future, over most land areas, the frequency of warm spells or heat waves would very likely increase. Other likely changes are listed below:

- Increased areas will be affected by drought

- There will be increased intense tropical cyclone activity

- There will be increased incidences of extreme high sea level (excluding tsunamis)

SHIFTING DISTRIBUTION OF SUMMER TEMPERATURE ANOMALIES

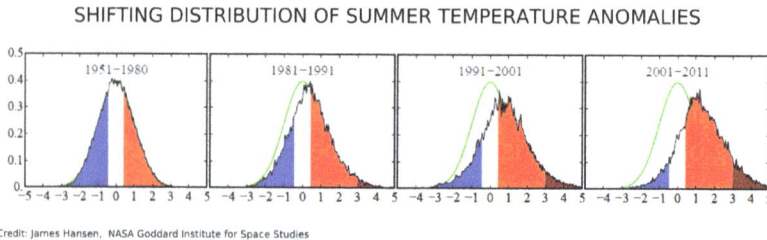

Credit: James Hansen, NASA Goddard Institute for Space Studies

Frequency of occurrence (vertical axis) of local June–July–August temperature anomalies (relative to 1951–1980 mean) for Northern Hemisphere land in units of local standard deviation (horizontal axis). According to Hansen *et al.* (2012), the distribution of anomalies has shifted to the right as a consequence of global warming, meaning that unusually hot summers have become more common. This is analogous to the rolling of a die: cool summers now cover only half of one side of a six-sided die, white covers one side, red covers four sides, and an extremely hot (red-brown) anomaly covers half of one side.

Storm strength leading to extreme weather is increasing, such as the power dissipation index of hurricane intensity. Kerry Emanuel writes that hurricane power dissipation is highly correlated with temperature, reflecting global warming. However, a further study by Emanuel using current model output concluded that the increase in power dissipation in recent decades cannot be completely attributed to global warming. Hurricane modeling has produced similar results, finding that hurricanes, simulated under warmer, high-CO_2 conditions, are more intense, however, hurricane frequency will be reduced. Worldwide, the proportion of hurricanes reaching categories 4 or 5 – with wind speeds above 56 metres per second – has risen from 20% in the 1970s to 35% in the 1990s. Precipitation hitting the US from hurricanes has increased by 7% over the 20th century. The extent to which this is due to global warming as opposed to the Atlantic Multidecadal Oscillation is unclear. Some studies have found that the increase in sea surface temperature may be offset by an increase in wind shear, leading to little or no change in hurricane activity. Hoyos *et al.* (2006) have linked the increasing trend in number of category 4 and 5 hurricanes for the period 1970–2004 directly to the trend in sea surface temperatures.

Thomas Knutson and Robert E. Tuleya of NOAA stated in 2004 that warming induced by greenhouse gas may lead to increasing occurrence of highly destructive category-5 storms. In 2008, Knutson *et al.* found that Atlantic hurricane and tropical storm frequencies could reduce under future greenhouse-gas-induced warming. Vecchi and Soden find that wind shear, the increase of which acts to inhibit tropical cyclones, also changes in model-projections of global warming. There are projected increases of wind shear in the tropical Atlantic and East Pacific associated with the deceleration of the Walker circulation, as well as decreases of wind shear in the western and central Pacific. The study does not make claims about the net effect on Atlantic and East Pacific hurricanes of the warming and moistening atmospheres, and the model-projected increases in Atlantic wind shear.

The World Meteorological Organization explains that "though there is evidence both for and against the existence of a detectable anthropogenic signal in the tropical cyclone climate record to date, no firm conclusion can be made on this point." They also clarified that "no individual tropical cyclone can be directly attributed to climate change."

A substantially higher risk of extreme weather does not necessarily mean a noticeably greater risk of slightly-above-average weather. However, the evidence is clear that severe weather and moderate rainfall are also increasing. Increases in temperature are expected to produce more intense convection over land and a higher frequency of the most severe storms.

Using the Palmer Drought Severity Index, a 2010 study by the National Center for Atmospheric Research projects increasingly dry conditions across much of the globe in the next 30 years, possibly reaching a scale in some regions by the end of the century that has rarely, if ever, been observed in modern times.

Coumou *et al.* (2013) estimated that global warming had increased the probability of local record-breaking monthly temperatures worldwide by a factor of 5. This was compared to a baseline climate in which no global warming had occurred. Using a medium global warming scenario, they project that by 2040, the number of monthly heat records globally could be more than 12 times greater than that of a scenario with no long-term warming.

Increased Evaporation

Over the course of the 20th century, evaporation rates have reduced worldwide; this is thought by many to be explained by global dimming. As the climate grows warmer and the causes of global dimming are reduced, evaporation will increase due to warmer oceans. Because the world is a closed system this will cause heavier rainfall, with more erosion. This erosion, in turn, can in vulnerable tropical areas (especially in Africa) lead to desertification. On the other hand, in other areas, increased rainfall lead to growth of forests in dry desert areas.

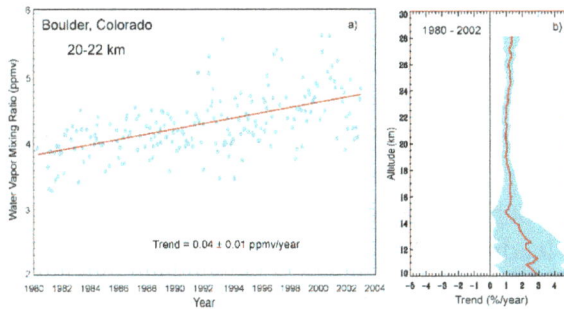

Increasing water vapor at Boulder, Colorado.

Scientists have found evidence that increased evaporation could result in more extreme weather as global warming progresses. The IPCC Third Annual Report says: "...global average water vapor concentration and precipitation are projected to increase during the 21st century. By the second half of the 21st century, it is likely that precipitation will have increased over northern mid- to high latitudes and Antarctica in winter. At low latitudes there are both regional increases and decreases over land areas. Larger year-to-year variations in precipitation are very likely over most areas where an increase in mean precipitation is projected."

Increased Freshwater Flow

Research based on satellite observations, published in October 2010, shows an increase in the flow of freshwater into the world's oceans, partly from melting ice and partly from increased precipitation driven by an increase in global ocean evaporation. The increase in global freshwater flow, based on data from 1994 to 2006, was about 18%. Much of the increase is in areas which already experience high rainfall. One effect, as perhaps experienced in the 2010 Pakistan floods, is to overwhelm flood control infrastructure.

Regional Climate Change

General Effects

In a literature assessment, Hegerl *et al.* (2007) assessed evidence for attributing observed climate change. They concluded that since the middle of the 20th century, it was likely that human influences had significantly contributed to surface temperature increases in every continent except Antarctica. The magazine *Scientific American* reported on December 23, 2008, that the 10 places most affected by climate change were Darfur, the Gulf Coast, Italy, northern Europe, the Great Barrier Reef, island nations, Washington, D.C., the Northwest Passage, the Alps, and Uganda.

Northern Hemisphere

In the northern hemisphere, the southern part of the Arctic region (home to 4,000,000 people) has experienced a temperature rise of 1 °C to 3 °C (1.8 °F to 5.4 °F) over the last 50 years. Canada, Alaska and Russia are experiencing initial melting of permafrost. This may disrupt ecosystems and by increasing bacterial activity in the soil lead to these areas becoming carbon sources instead of carbon sinks. A study (published in *Science*) of changes to eastern Siberia's permafrost suggests that it is gradually disappearing in the southern regions, leading to the loss of nearly 11% of Siberia's nearly 11,000 lakes since 1971. At the same time, western Siberia is at the initial stage where melting permafrost is creating new lakes, which will eventually start disappearing as in the east. Furthermore, permafrost melting will eventually cause methane release from melting permafrost peat bogs.

Polar Regions

Anisimov *et al.* (2007) assessed the literature on impacts of climate change in Polar regions. Model projections showed that Arctic terrestrial ecosystems and the active layer (the top layer of soil or rock in permafrost that is subjected to seasonal freezing and thawing) would be a small sink for carbon (i.e., net uptake of carbon) over this century (p. 662). These projections were viewed as being uncertain. It was judged that increased emissions of carbon from thawing of permafrost could occur. This would lead to an amplification of warming.

Atmosphere

Temperature trends in the lower stratosphere, mid to upper troposphere, lower troposphere, and surface, 1957-2005.

An enhanced greenhouse effect is expected to cause cooling in higher parts of the atmosphere. Cooling of the lower stratosphere (about 49,000-79,500 ft.) since 1979 is shown by both satellite Microwave sounding unit and radiosonde data, but is larger in the radiosonde data likely due to uncorrected errors in the radiosonde data.

A contraction of the thermosphere has been observed as a possible result in part due to increased carbon dioxide concentrations, the strongest cooling and contraction occurring in that layer during solar minimum. The most recent contraction in 2008–2009 was the largest such since at least 1967.

Recent evidence suggests that warming of the tropical oceans since a "tipping point" in 2000 may have acted as a negative feedback, reducing the observed warming during the 2000s (decade). As warming and evaporation above the Pacific Ocean, temperatures in the lower stratosphere near the tropopause declined due to both greenhouse gases and ozone-depleting substances, reducing water vapor levels and removing its warming effect, with vapor concentrations below 2.2 ppmv as measured by the HALOE instrument on the Upper Atmosphere Research Satellite, in the lower stratosphere of the tropics between 5°N - 5°S first being observed since 2001, although a reversal in this pattern is also likely. The water vapor in the stratosphere arrives through tall thunderstorms, while 15% of this vapor is delivered by tropical cyclones, and through chemical breakdown of methane into water vapor and carbon dioxide, both of which are greenhouse gases. The vapor is frozen out of the stratosphere as more of the water is subjected to temperatures that freeze it out of the stratosphere. Water vapor concentrations in the lower stratosphere have declined by 10% (0.4 ppmv) since 2000, reducing warming during the decade by 25%. A rapid cooling of 4 °C to 6 °C also occurred in the lower stratosphere in the mid-1990s, while the rate of ocean warming increased. During the 1990s, increased stratospheric water vapor led to a 30% increase in warming. After 2000, the sea surface temperatures of the tropical Western Pacific, where a warm pool of water exists and where temperatures are heavily influenced by

ENSO, between 10°N - 10°S and 139° - 171° longitude became anti-correlated with temperatures at the tropopause in the same latitudes between 171° - 200° longitude, both measured since the early 1980s; although the correlation had been previously positive, since 2000 the SST anomalies increased while tropopause temperatures decreased. A sharp increase in average SSTs within the Western Pacific warm pool by more than 0.25 °C in 2000, which has since stabilized, occurred as the "cold point" temperature of the study area at the tropopause experienced a significant reduction. This resulted in less water vapor from tropical thunderstorms entering the stratosphere. However, prior to 2000, increases in average Western Pacific SSTs had resulted in increases in tropopause cold point temperatures.

Geophysical Systems

Biogeochemical Cycles

Climate change can have an effect on the carbon cycle in an interactive "feedback" process . A feedback exists where an initial process triggers changes in a second process that in turn influences the initial process. A positive feedback intensifies the original process, and a negative feedback reduces it (IPCC, 2007d:78). Models suggest that the interaction of the climate system and the carbon cycle is one where the feedback effect is positive (Schneider *et al.*, 2007).

Using the A2 SRES emissions scenario, Schneider *et al.* (2007) found that this effect led to additional warming by 2100, relative to the 1990–2000 period, of 0.1 to 1.5 °C. This estimate was made with high confidence. The climate projections made in the IPCC Fourth Assessment Report of 1.1 to 6.4 °C account for this feedback effect. On the other hand, with medium confidence, Schneider *et al.* (2007) commented that additional releases of GHGs were possible from permafrost, peat lands, wetlands, and large stores of marine hydrates at high latitudes.

Gas Hydrates

Gas hydrates are ice-like deposits containing a mixture of water and gas, the most common gas of which is methane (Maslin, 2004:1). Gas hydrates are stable under high pressures and at relatively low temperatures and are found underneath the oceans and permafrost regions. Future warming at intermediate depths in the world's oceans, as predicted by climate models, will tend to destabilize gas hydrates resulting in the release of large quantities of methane. On the other hand, projected rapid sea level rise in the coming centuries associated with global warming will tend to stabilize marine gas hydrate deposits.

Carbon Cycle

Models have been used to assess the effect that climate change will have on the carbon cycle (Meehl *et al.*, 2007-790). In the Coupled Climate-Carbon Cycle Model Intercomparison Project, eleven climate models were used. Observed emissions were used in the models and future emission projections were based on the IPCC SRES A2 emissions scenario.

Unanimous agreement was found among the models that future climate change will reduce the efficiency of the land and ocean carbon cycle to absorb human-induced CO_2. As a result, a larger fraction of human-induced CO_2 will stay airborne if climate change controls the carbon cycle. By the end of the 21st century, this additional CO_2 in the atmosphere varied between 20 and 220 ppm for the two extreme models, with most models lying between 50 and 100 ppm. This additional CO_2 led to a projected increase in warming of between 0.1 and 1.5 °C.

Cryosphere

Northern Hemisphere average annual snow cover has declined in recent decades. This pattern is consistent with warmer global temperatures. Some of the largest declines have been observed in the spring and summer months.

Sea Ice

As the climate warms, snow cover and sea ice extent decrease. Large-scale measurements of sea-ice have only been possible since the satellite era, but through looking at a number of different satellite estimates, it has been determined that September Arctic sea ice has decreased between 1973 and 2007 at a rate of about -10% +/- 0.3% per decade. Sea ice extent for September for 2012 was by far the lowest on record at 3.29 million square kilometers, eclipsing the previous record low sea ice extent of 2007 by 18%. The age of the sea ice is also an important feature of the state of the sea ice cover, and for the month of March 2012, older ice (4 years and older) has decreased from 26% of the ice cover in 1988 to 7% in 2012. Sea ice in the Antarctic has shown very little trend over the same period, or even a slight increase since 1979. Though extending the Antarctic sea-ice record back in time is more difficult due to the lack of direct observations in this part of the world.

In a literature assessment, Meehl *et al.* (2007:750) found that model projections for the 21st century showed a reduction of sea ice in both the Arctic and Antarctic. The range of model responses was large. Projected reductions were accelerated in the Arctic. Using the high-emission A2 SRES scenario, some models projected that summer sea ice cover in the Arctic would disappear entirely by the latter part of the 21st century.

Glacier Retreat and Disappearance

Warming temperatures lead to the melting of glaciers and ice sheets. IPCC (2007a:5) found that, on average, mountain glaciers and snow cover had decreased in both the northern and southern hemispheres. This widespread decrease in glaciers and ice caps has contributed to observed sea level rise.

As stated above, the total volume of glaciers on Earth is declining sharply. Glaciers have been retreating worldwide for at least the last century; the rate of retreat has increased in the past decade. Only a few glaciers are actually advancing (in locations that were well below freezing, and where increased precipitation has outpaced melting). The progressive disappearance of glaciers has implications not only for a rising global sea level, but also for water supplies in certain regions of Asia and South America.

With very high or high confidence, IPCC (2007d:11) made a number of projections related to future changes in glaciers:

- Mountainous areas in Europe will face glacier retreat

- In Latin America, changes in precipitation patterns and the disappearance of glaciers will significantly affect water availability for human consumption, agriculture, and energy production

- In Polar regions, there will be reductions in glacier extent and the thickness of glaciers.

In historic times, glaciers grew during a cool period from about 1550 to 1850 known as the Little Ice Age. Subsequently, until about 1940, glaciers around the world retreated as the climate warmed. Glacier retreat declined and reversed in many cases from 1950 to 1980 as a slight global cooling occurred. Since 1980, glacier retreat has become increasingly rapid and ubiquitous, and has threatened the existence of many of the glaciers of the world. This process has increased markedly since 1995.

Excluding the ice caps and ice sheets of the Arctic and Antarctic, the total surface area of glaciers worldwide has decreased by 50% since the end of the 19th century. Currently glacier retreat rates and mass balance losses have been increasing in the Andes, Alps, Pyrenees, Himalayas, Rocky Mountains and North Cascades.

The loss of glaciers not only directly causes landslides, flash floods and glacial lake overflow, but also increases annual variation in water flows in rivers. Glacier runoff declines in the summer as glaciers decrease in size, this decline is already observable in several regions. Glaciers retain water on mountains in high precipitation years, since the snow cover accumulating on glaciers protects the ice from melting. In warmer and drier years, glaciers offset the lower precipitation amounts with a higher meltwater input.

Of particular importance are the Hindu Kush and Himalayan glacial melts that comprise the principal dry-season water source of many of the major rivers of the Central, South, East and Southeast Asian mainland. Increased melting would cause greater flow for several decades, after which "some areas of the most populated regions on Earth are likely to 'run out of water'" as source glaciers are depleted. The Tibetan Plateau contains the world's third-largest store of ice. Temperatures there are rising four times faster than in the rest of China, and glacial retreat is at a high speed compared to elsewhere in the world.

According to a Reuters report, the Himalayan glaciers that are the sources of Asia's biggest rivers—Ganges, Indus, Brahmaputra, Yangtze, Mekong, Salween and Yellow—could diminish as temperatures rise. Approximately 2.4 billion people live in the drainage basin of the Himalayan rivers. India, China, Pakistan, Bangladesh, Nepal and Myanmar could experience floods followed by droughts in coming decades. The Indus, Ganges and Brahmaputra river basins support 700 million people in Asia. In India alone, the Ganges provides water for drinking and farming for more than 500 million people. It has to be acknowledged, however, that increased seasonal runoff of Himalayan glaciers led to increased agricultural production in northern India throughout the 20th century. Research studies suggest that climate change will have a marked affect on meltwater in the Indus Basin.

The recession of mountain glaciers, notably in Western North America, Franz-Josef Land, Asia, the Alps, the Pyrenees, Indonesia and Africa, and tropical and sub-tropical regions of South America, has been used to provide qualitative support to the rise in global temperatures since the late 19th century. Many glaciers are being lost to melting further raising concerns about future local water resources in these glaciated areas. In Western North America the 47 North Cascade glaciers observed all are retreating.

Despite their proximity and importance to human populations, the mountain and valley glaciers of temperate latitudes amount to a small fraction of glacial ice on the earth. About 99% is in the great ice sheets of polar and subpolar Antarctica and Greenland. These continuous continental-scale ice sheets, 3 kilometres (1.9 mi) or more in thickness, cap the polar and subpolar land masses. Like rivers flowing from an enormous lake, numerous outlet glaciers transport ice from the margins of the ice sheet to the ocean.

Glacier retreat has been observed in these outlet glaciers, resulting in an increase of the ice flow rate. In Greenland the period since the year 2000 has brought retreat to several very large glaciers that had long been stable. Three glaciers that have been researched, Helheim, Jakobshavn Isbræ and Kangerdlugssuaq Glaciers, jointly drain more than 16% of the Greenland Ice Sheet. Satellite images and aerial photographs from the 1950s and 1970s show that the front of the glacier had remained in the same place for decades. But in 2001 it began retreating rapidly, retreating 7.2 km (4.5 mi) between 2001 and 2005. It has also accelerated from 20 m (66 ft)/day to 32 m (105 ft)/day. Jakobshavn Isbræ in western Greenland had been moving at speeds of over 24 m (79 ft)/day with a stable terminus since at least 1950. The glacier's ice tongue began to break apart in 2000, leading to almost complete disintegration in 2003, while the retreat rate increased to over 30 m (98 ft)/day.

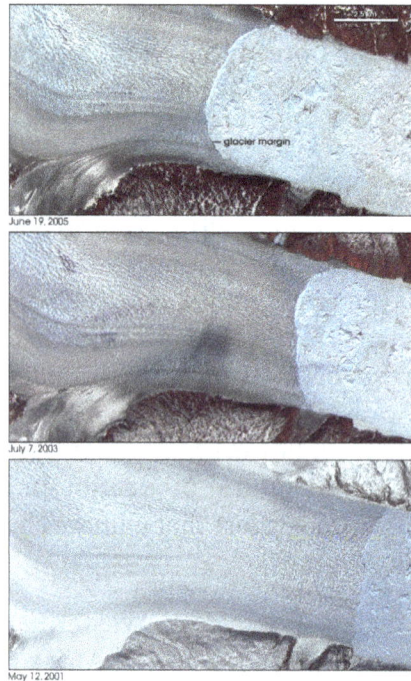

Retreat of the Helheim Glacier, Greenland

Oceans

The role of the oceans in global warming is a complex one. The oceans serve as a sink for carbon dioxide, taking up much that would otherwise remain in the atmosphere, but increased levels of CO_2 have led to ocean acidification. Furthermore, as the temperature of the oceans increases, they become less able to absorb excess CO_2. Global warming is projected to have a number of effects on the oceans. Ongoing effects include rising sea levels due to thermal expansion and melting of glaciers and ice sheets, and warming of the ocean surface, leading to increased temperature stratification. Other possible effects include large-scale changes in ocean circulation.

Sea Level Rise

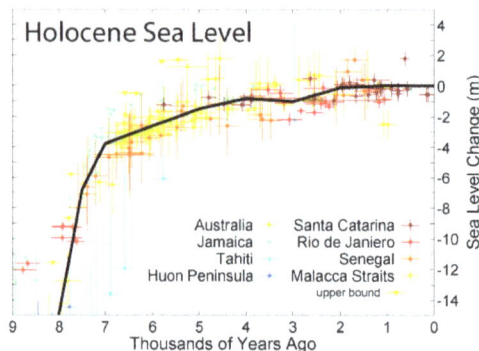

Sea level rise during the Holocene.

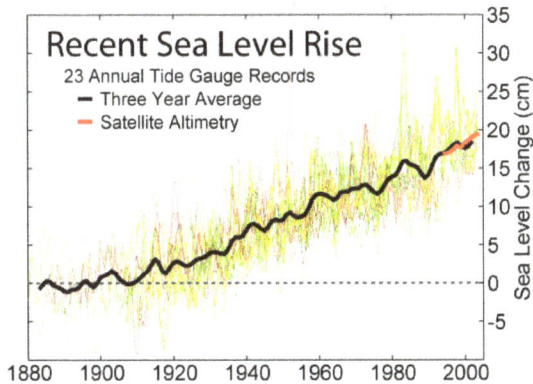

Sea level has been rising 0.2 cm/year, based on measurements of sea level rise from 23 long tide gauge records in geologically stable environments.

IPCC (2007a:5) reported that since 1961, global average sea level had risen at an average rate of 1.8 [1.3 to 2.3] mm/yr. Between 1993 and 2003, the rate increased above the previous period to 3.1 [2.4 to 3.8] mm/yr. IPCC (2007a) were uncertain whether the increase in rate from 1993 to 2003 was due to natural variations in sea level over the time period, or whether it reflected an increase in the underlying long-term trend.

IPCC (2007a:13, 14) projected sea level rise to the end of the 21st century using the SRES emission scenarios. Across the six SRES marker scenarios, sea level was projected to rise by 18 to 59 cm (7.1 to 23.2 inches). This projection was for the time period 2090–2099, with the increase in level relative to average sea levels over the 1980–1999 period. Due to a lack of scientific understanding, this sea level rise estimate does not include all of the possible contributions of ice sheets.

With increasing average global temperature, the water in the oceans expands in volume, and additional water enters them which had previously been locked up on land in glaciers and ice sheets. The Greenland and the Antarctic ice sheets are major ice masses, and at least the former of which may suffer irreversible decline. For most glaciers worldwide, an average volume loss of 60% until 2050 is predicted. Meanwhile, the estimated total ice melting rate over Greenland is 239 ± 23 cubic kilometres (57.3 ± 5.5 cu mi) per year, mostly from East Greenland. The Antarctic ice sheet, however, is expected to grow during the 21st century because of increased precipitation. Under the IPCC Special Report on Emission Scenario (SRES) A1B, by the mid-2090s global sea level will reach 0.22 to 0.44 m (8.7 to 17.3 in) above 1990 levels, and is currently rising at about 4 mm (0.16 in) per year. Since 1900, the sea level has risen at an average of 1.7 mm (0.067 in) per year; since 1993, satellite altimetry from TOPEX/Poseidon indicates a rate of about 3 mm (0.12 in) per year.

The sea level has risen more than 120 metres (390 ft) since the Last Glacial Maximum about 20,000 years ago. The bulk of that occurred before 7000 years ago. Global temperature declined after the Holocene Climatic Optimum, causing a sea level lowering of 0.7 ± 0.1 m (27.6 ± 3.9 in) between 4000 and 2500 years before present. From 3000 years ago to the start of the 19th century, sea level was almost constant, with only minor fluctuations. However, the Medieval Warm Period may have caused some sea level rise; evidence has been found in the Pacific Ocean for a rise to perhaps 0.9 m (2 ft 11 in) above present level in 700 BP.

In a paper published in 2007, the climatologist James E. Hansen *et al.* claimed that ice at the poles does not melt in a gradual and linear fashion, but that another according to the geological record, the ice sheets can suddenly destabilize when a certain threshold is exceeded. In this paper Hansen *et al.* state:

Our concern that BAU GHG scenarios would cause large sealevel rise this century (Hansen 2005) differs from estimates of IPCC (2001, 2007), which foresees little or no contribution to twentyfirst century sealevel rise from Greenland and Antarctica. However, the IPCC analyses and projections do not well account for the nonlinear physics of wet ice sheet disintegration, ice streams and eroding ice shelves, nor are they consistent with the palaeoclimate evidence we have presented for the absence of discernible lag between ice sheet forcing and sealevel rise.

Sea level rise due to the collapse of an ice sheet would be distributed nonuniformly across the globe. The loss of mass in the region around the ice sheet would decrease the gravitational potential there, reducing the amount of local sea level rise or even causing local sea level fall. The loss of the localized mass would also change the moment of inertia of the Earth, as flow in the Earth's mantle will require 10–15 thousand years to make up the mass deficit. This change in the moment of inertia results in true polar wander, in which the Earth's rotational axis remains fixed with respect to the sun, but the rigid sphere of the Earth rotates with respect to it. This changes the location of the equatorial bulge of the Earth and further affects the geoid, or global potential field. A 2009 study of the effects of collapse of the West Antarctic Ice Sheet shows the result of both of these effects. Instead of a global 5-meter sea level rise, western Antarctica would experience approximately 25 centimeters of sea level fall, while the United States, parts of Canada, and the Indian Ocean, would experience up to 6.5 meters of sea level rise.

A paper published in 2008 by a group of researchers at the University of Wisconsin led by Anders Carlson used the deglaciation of North America at 9000 years before present as an analogue to predict sea level rise of 1.3 meters in the next century, which is also much higher than the IPCC projections. However, models of glacial flow in the smaller present-day ice sheets show that a probable maximum value for sea level rise in the next century is 80 centimeters, based on limitations on how quickly ice can flow below the equilibrium line altitude and to the sea.

Temperature Rise and Ocean Heat Content

Time series of seasonal (red dots) and annual average (black line) global upper ocean heat content for the 0-700m layer between 1955 and 2008. The graph shows that ocean heat content has increased over this time period.

From 1961 to 2003, the global ocean temperature has risen by 0.10 °C from the surface to a depth of 700 m. There is variability both year-to-year and over longer time scales, with global ocean heat content observations showing high rates of warming for 1991 to 2003, but some cooling from 2003 to 2007. Nevertheless, there is a strong trend during the period of reliable measurements. Increasing heat content in the ocean is also consistent with sea level rise, which is occurring mostly as a result of thermal expansion of the ocean water as it warms.

The temperature of the Antarctic Southern Ocean rose by 0.17 °C (0.31 °F) between the 1950s and the 1980s, nearly twice the rate for the world's oceans as a whole. As well as having effects on ecosystems (e.g. by melting sea ice, affecting algae that grow on its underside), warming reduces the ocean's ability to absorb CO_2.

Acidification

Ocean acidification is an effect of rising concentrations of CO_2 in the atmosphere, and is not a direct consequence of global warming. The oceans soak up much of the CO_2 produced by living organisms, either as dissolved gas, or in the skeletons of tiny marine creatures that fall to the bottom to become chalk or limestone. Oceans currently absorb about one tonne of CO_2 per person per year. It is estimated that the oceans have absorbed around half of all CO_2 generated by human activities since 1800 (118 ± 19 petagrams of carbon from 1800 to 1994).

In water, CO_2 becomes a weak carbonic acid, and the increase in the greenhouse gas since the Industrial Revolution has already lowered the average pH (the laboratory measure of acidity) of seawater by 0.1 units, to 8.2. Predicted emissions could lower the pH by a further 0.5 by 2100, to a level probably not seen for hundreds of millennia and, critically, at a rate of change probably 100 times greater than at any time over this period.

There are concerns that increasing acidification could have a particularly detrimental effect on corals (16% of the world's coral reefs have died from bleaching caused by warm water in 1998, which coincidentally was, at the time, the warmest year ever recorded) and other marine organisms with calcium carbonate shells.

In November 2009 an article in *Science* by scientists at Canada's Department of Fisheries and Oceans reported they had found very low levels of the building blocks for the calcium chloride that forms plankton shells in the Beaufort Sea. Fiona McLaughlin, one of the DFO authors, asserted that the increasing acidification of the Arctic Ocean was close to the point it would start dissolving the walls of existing plankton: *"[the] Arctic ecosystem may be risk. In actual fact, they'll dissolve the shells."* Because cold water absorbs CO_2 more readily than warmer water the acidification is more severe in the polar regions. McLaughlin predicted the acidified water would travel to the North Atlantic within the next ten years.

Shutdown of Thermohaline Circulation

There is some speculation that global warming could, via a shutdown or slowdown of the thermohaline circulation, trigger localized cooling in the North Atlantic and lead to cooling, or lesser warming, in that region. This would affect in particular areas like Scandinavia and Britain that are warmed by the North Atlantic drift.

The chances of this near-term collapse of the circulation are unclear; there is some evidence for the short-term stability of the Gulf Stream and possible weakening of the North Atlantic drift. However, the degree of weakening, and whether it will be sufficient to shut down the circulation, is under debate. As yet, no cooling has been found in northern Europe or nearby seas. Lenton et al. found that "simulations clearly pass a THC tipping point this century".

IPCC (2007b:17) concluded that a slowing of the Meridional Overturning Circulation would very likely occur this century. Due to global warming, temperatures across the Atlantic and Europe were still projected to increase.

Oxygen Depletion

The amount of oxygen dissolved in the oceans may decline, with adverse consequences for ocean life.

Sulfur Aerosols

Sulfur aerosols, especially stratospheric sulfur aerosols have a significant effect on climate. One source of such aerosols is the sulfur cycle, where plankton release gases such as DMS which eventually becomes oxidised to sulfur dioxide in the atmosphere. Disruption to the oceans as a result of ocean acidification or disruptions to the thermohaline circulation may result in disruption of the sulfur cycle, thus reducing its cooling effect on the planet through the creation of stratospheric sulfur aerosols.

Geology

Volcanoes

The retreat of glaciers and ice caps can cause increased volcanism. Reduction in ice cover reduces the confining pressure exerted on the volcano, increasing deviatoric stresses and potentially causing the volcano to erupt. This reduction of pressure can also cause decompression melting of material in the mantle, resulting in the generation of more magma. Researchers in Iceland have shown that the rate of volcanic rock production there following deglaciation (10,000 to 4500 years before present) was 20–30 times greater than that observed after 2900 years before present. While the original study addresses the first reason for increased volcanism (reduced confining pressure), scientists have more recently shown that these lavas have unusually high trace element concentrations, indicative of increased melting in the mantle. This work in Iceland has been corroborated by a study in California, in which scientists found a strong correlation between volcanism and periods of global deglaciation. The effects of current sea level rise could include increased crustal stress at the base of coastal volcanoes from a rise in the volcano's water table (and the associated saltwater intrusion), while the mass from extra water could activate dormant seismic faults around volcanoes. In addition, the wide-scale displacement of water from melting in places such as West Antarctica is likely to slightly alter the Earth's rotational period and may shift its axial tilt on the scale of hundreds of metres, inducing further crustal stress changes.

Current melting of ice is predicted to increase the size and frequency of volcanic eruptions. In particular, lateral collapse events at stratovolcanoes are likely to increase, and there are potential positive feedbacks between the removal of ice and magmatism.

Earthquakes

A numerical modeling study has demonstrated that seismicity increases during unloading, such as that due to the removal of ice.

Climate of Antarctica

The climate of Antarctica is the coldest on Earth. Antarctica's lowest air temperature record was set on 21 July 1983, with −89.2 °C (−128.6 °F) at Vostok Station. Satellite measurements have identified even lower ground temperatures, down to −93.2 °C (−135.8 °F) at the cloud free East Antarctic Plateau on 10 August 2010. It is also extremely dry (technically a desert), averaging 166 mm (6.5 in) of precipitation per year. On most parts of the continent the snow rarely melts and is eventually compressed to become the glacier ice that makes up the ice sheet. Weather fronts rarely penetrate far into the continent, because of the katabatic winds. Most of Antarctica has an ice cap climate (Köppen *EF*) with very cold, generally extremely dry weather.

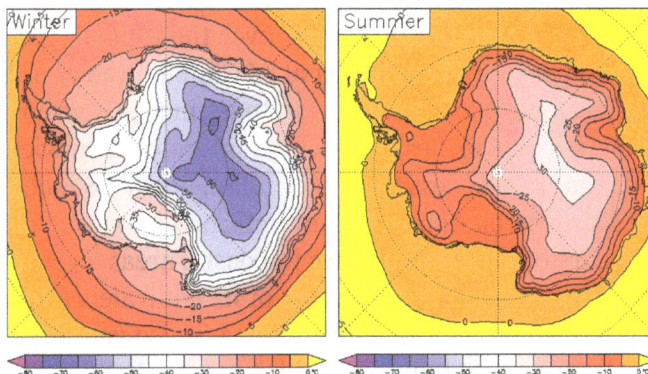

Surface temperature of Antarctica in winter and summer from the
European Centre for Medium-Range Weather Forecasts

Temperature

The lowest reliably measured temperature of a continuously occupied station on Earth of −89.2 °C (−128.6 °F) was on 21 July 1983 at Vostok Station. For comparison, this is 10.7 °C (19.3 °F) colder than subliming dry ice (at sea level pressure). The altitude of the location is 3,900 meters (12,800 feet).

The lowest recorded temperature of any location on Earth's surface was −93.2 °C (−135.8 °F) at 81°48′S 63°30′E81.8°S 63.5°E, which is on an unnamed Antarctic plateau between Dome A and Dome F, on August 10, 2010. The temperature was deduced from radiance measured by the Landsat 8 satellite, and discovered during a National Snow and Ice Data Center review of stored data in December, 2013. This temperature is not directly comparable to the -89.2 quoted above, since it is a skin temperature deduced from satellite-measured upwelling radiance, rather than a thermometer-measured temperature of the air 1.5 m (4.9 ft) above the ground surface.

On the coast Antarctic average temperatures are around -10c (in the warmest parts of Antarctica) and in the elevated inland they average about -55c in Vostok.

The highest temperature ever recorded in Antarctica was 17.5 °C (63.5 °F) at Esperanza Base, on the Antarctic Peninsula, on 24 March 2015. The mean annual temperature of the interior is −57 °C (−70.6 °F). The coast is warmer. Monthly means at McMurdo Station range from −26 °C (−14.8 °F) in August to −3 °C (26.6 °F) in January. At the South Pole, the highest temperature ever recorded was −12.3 °C (9.9 °F) on 25 December 2011. Along the Antarctic Peninsula, temperatures as high as 15 °C (59 °F) have been recorded, though the summer temperature is below 0 °C (32 °F) most of the time. Severe low temperatures vary with latitude, elevation, and distance from the ocean. East Antarctica is colder than West Antarctica because of its higher elevation. The Antarctic Peninsula has the most moderate climate. Higher temperatures occur in January along the coast and average slightly below freezing.

Precipitation

The total precipitation on Antarctica, averaged over the entire continent, is about 166 millimetres (6.5 inches) per year (Vaughan et al., J Climate, 1999). The actual rates vary widely, from high values over the Peninsula (15 to 25 inches a year) to very low values (as little as 50 millimetres (2.0 inches) in the high interior (Bromwich, Reviews of Geophysics, 1988). Areas that receive less than 250 millimetres (9.8 inches) of precipitation per year are classified as deserts. Almost all Antarctic precipitation falls as snow. Rainfall is rare and mainly occurs during the summer in coastal areas and surrounding islands. Note that the quoted precipitation is a measure of its equivalence to water, rather than being the actual depth of snow. The air in Antarctica is also very dry. The low temperatures result in a very low absolute humidity, which means that dry skin and cracked lips are a continual problem for scientists and expeditioners working in the continent.

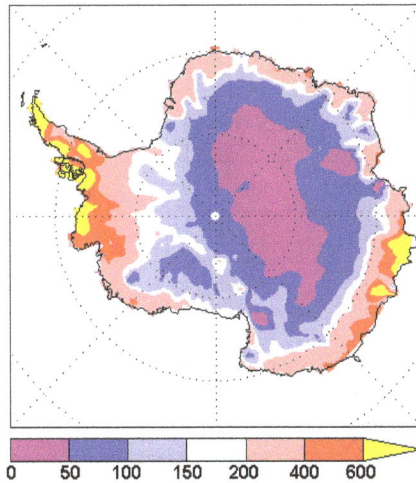

Map of average annual precipitation on Antarctica (mm liquid equivalent).

Weather Condition Classification

The weather in Antarctica can be highly variable, and the weather conditions can often change dramatically in short periods of time. There are various classifications for describing weather conditions in Antarctica; restrictions given to workers during the different conditions vary by station and nation. In Antarctic, there are different stations scattered everywhere in Antarctica, 16 total in Antarctica (Amery, Burger Hills, Cape Poinsett, Casey, Davis, Dumont D'urville, Haupt Nunatak, Law Dome, Mawson, McMurdo, Mirnyj, Novolazarevskaja, Skiway South, Syowa, Whoop Whoop and Wilkins Runway) and one in sub-Antarctica (Macquarie Island). The three places that have the lowest temperature are Amery, Law Dome and Wilkins Runway (temperature in order: -24.1, -12.8,-10.2).

Ice Cover

Nearly all of Antarctica is covered by a sheet of ice that is, on average, a mile thick or more (1.6 km). Antarctica contains 90% of the world's ice and more than 70% of its fresh water. If all the land-ice covering Antarctica were to melt — around 30 million cubic kilometres (7.2 million cubic miles) of ice — the seas would rise by over 60 metres (200 feet). This is, however, very unlikely within the next few centuries. The Antarctic is so cold that even with increases of a few degrees, temperatures would generally remain below the melting point of ice. Higher temperatures are expected to lead to more snow, which would increase the amount of ice in Antarctica, offsetting approximately one third of the expected sea level rise from thermal expansion of the oceans. During a recent decade, East Antarctica thickened at an average rate of about 1.8 centimetres per year while West Antarctica showed an overall thinning of 0.9 centimetres per year. Because ice flows, albeit slowly, the ice within the ice sheet is younger than the age of the sheet itself.

Morphometric data for Antarctica (from Drewry, 1983)					
Surface	**Area (km²)**	**Per-cent**	**Mean ice thickness (m)**	**Volume (km³)**	**Percent**
Inland ice sheet	11,965,700	85.97	2,450	29,324,700	97.00
Ice shelves	1,541,710	11.08	475	731,900	2.43
Ice rises	78,970	.57	670	53,100	.18
Glacier ice (total)	13,586,380		2,160	30,109,800[1]	
Rock outcrop	331,690	2.38			
Antarctica (total)	13,918,070	100.00	2,160	30,109,800[1]	100.00

[1]The total ice volume is different from the sum of the component parts because individual figures have been rounded.

Regional ice data (from Drewry and others, 1982; Drewry, 1983)			
Region	**Area (km²)**	**Mean ice thickness (m)**	**Volume (km³)**
East Antarctica			
Inland ice	9,855,570	2,630	25,920,100
Ice shelves	293,510	400	117,400
Ice rises	4,090	400	1,600
West Antarctica (excluding Antarctic Peninsula)			
Inland ice sheet	1,809,760	1,780	3,221,400
Ice shelves	104,860	375	39,300
Ice rises	3,550	375	1,300
Antarctic Peninsula			
Inland ice sheet	300,380	610	183,200
Ice shelves	144,750	300	43,400
Ice rises	1,570	300	500
Ross Ice Shelf			
Ice shelf	525,840	427	224,500
Ice rises	10,320	500	5,100
Filchner-Ronne Ice Shelf			
Ice shelf	472,760	650	307,300
Ice rises	59,440	750	44,600

Ice Shelves

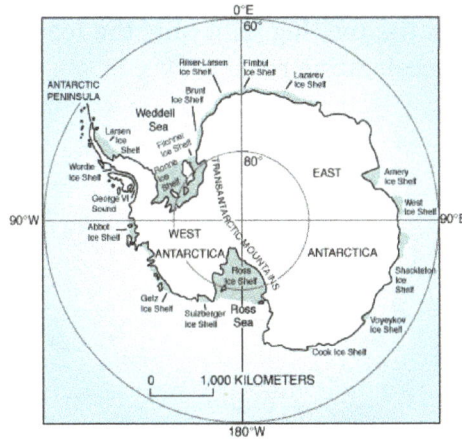

Antarctic ice shelves, 1998

About 75% of the coastline of Antarctica is ice shelves. The utmost parts consist of floating ice until the grounding line of land based glaciers is reached, which is determined through affords such as Operation IceBridge. Ice shelves lose mass through iceberg breakup (calving), or basal melting (at the foot of the glacier, when warm ocean water impacts), and this can affect ice sheet stability when the land based glaciers start to retreat; melting or breakup of floating shelf ice does not directly affect global sea levels, however, when sea ice freezes, it preferentially expels salt, in the process becoming purer than the sea water it is floating in. Pure water is less dense than salty water, so when the ice melts it will overflow the 'hole in the water' that the ice had occupied, and when it overflows, it raises the water level.

Known changes in coastline ice:

- Around the Antarctic Peninsula:

    o   1936–1989: Wordie Ice Shelf significantly reduced in size.

    o   1995: Ice in the Prince Gustav Channel disintegrated.

    o   Parts of the Larsen Ice Shelf broke up in recent decades.

- 1995: The Larsen A ice shelf disintegrated in January 1995.

- 2001: 3,250 square kilometres (1,250 square miles) of the Larsen B ice shelf disintegrated in February 2001. It had been gradually retreating before the breakup event.

- 2015: A study concluded that the remaining *Larsen B* ice-shelf will disintegrate by the end of the decade, based on observations of faster flow and rapid thinning of glaciers in the area.

The George VI Ice Shelf, which may be on the brink of instability, has probably existed for approximately 8,000 years, after melting 1,500 years earlier. Warm ocean currents may have been the cause of the melting. Not only the ice sheets are losing mass, but they are losing mass at an accelerating rate.

Global Warming

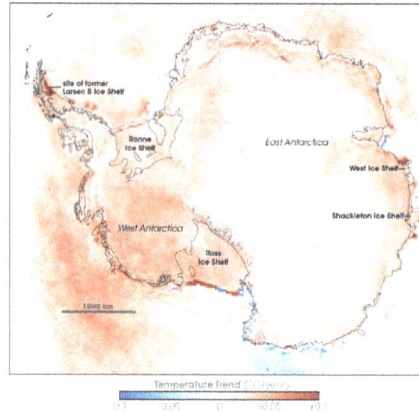

Antarctic Skin Temperature Trends between 1981 and 2007, based on thermal infrared observations made by a series of NOAA satellite sensors. Skin temperature trends do not necessarily reflect air temperature trends.

The continent-wide average surface temperature trend of Antarctica is positive and significant at >0.05 °C/decade since 1957. The West Antarctic ice sheet has warmed by more than 0.1 °C/decade in the last 50 years, and is strongest in winter and spring. Although this is partly offset by fall cooling in East Antarctica, this effect is restricted to the 1980s and 1990s.

Research published in 2009 found that overall the continent had become warmer since the 1950s, a finding consistent with the influence of man-made climate change:

> *"We can't pin it down, but it certainly is consistent with the influence of greenhouse gases from fossil fuels"*, said NASA scientist Drew Shindell, another study co-author. Some of the effects also could be natural variability, he said.

The British Antarctic Survey, which has undertaken the majority of Britain's scientific research in the area, stated in 2009:

- West Antarctic ice loss could contribute to 1.4 metres (4 feet 7 inches) sea level rise

- Antarctica predicted to warm by around 3 °C (5.4 °F) over this century

- 10% increase in sea ice around the Antarctic

- Rapid ice loss in parts of the Antarctic

- Warming of the Southern Ocean will cause changes in Antarctic ecosystem

- Hole in ozone layer, which has shielded most of Antarctica from global warming

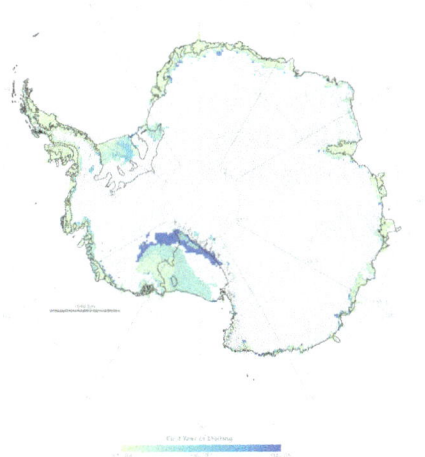

20 September 2007 NASA map showing previously un-melted snowmelt

The area of strongest cooling appears at the South Pole, and the region of strongest warming lies along the Antarctic Peninsula. A possible explanation is that loss of UV-absorbing ozone may have cooled the stratosphere and strengthened the polar vortex, a pattern of spinning winds around the South Pole. The vortex acts like an atmospheric barrier, preventing warmer, coastal air from moving into the continent's interior. A stronger polar vortex might explain the cooling trend in the interior of Antarctica.

In their latest study (20 September 2007) NASA researchers have confirmed that Antarctic snow is melting farther inland from the coast over time, melting at higher altitudes than ever and increasingly melting on Antarctica's largest ice shelf.

There is also evidence for widespread glacier retreat around the Antarctic Peninsula.

Researchers reported on 21 December 2012 in Nature Geoscience that from 1958 to 2010, the average temperature at the mile-high Byrd Station rose by 2.4 °C (4.3 °F), with warming fastest in its winter and spring. The spot which is in the heart of the West Antarctic Ice Sheet is one of the fastest-warming places on Earth. In 2015, the temperature showed changes but in a stable manner and the only months that have drastic change in that year are August and September. It also did show that the temperature was very stable throughout the year.

Antarctica Cooling Controversy

An apparent contradiction in the observed cooling behavior of Antarctica between 1966 and 2000 became part of the public debate in the global warming controversy, particularly between advocacy groups of both sides in the public arena including politicians, as well as the popular media. In his novel *State of Fear*, Michael Crichton asserted that the Antarctic data contradict global warming. The few scientists who have commented on the supposed controversy state that there is no contradiction, while the author of the paper whose work inspired Crichton's remarks has said that Crichton "misused" his results. There is no similar controversy within the scientific community, as the small observed changes in Antarctica are consistent with the small changes predicted by climate models, and because the overall trend since comprehensive observations began is now known to be one of warming. At the South Pole, where some of the strongest cooling trends were observed between the 1950s and 1990s, the mean trend is flat from 1957 through 2013.

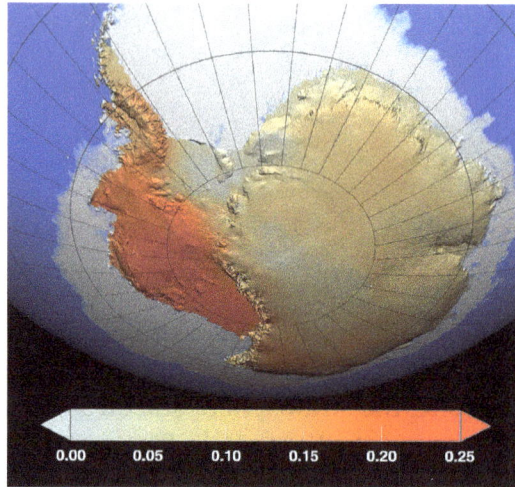

Antarctic surface temperature trends. Red represents areas where temperatures have increased
the most during the last 50 years, particularly in West Antarctica.
The temperature trends are given in °C/decade.

Background

Changes in the average temperature of the Antarctic continent has been the subject of various measurements. The trend differs at different locations on the continent. These trends have been labelled as "contradictory" in some accounts. Observations unambiguously show the Antarctic Peninsula to be warming. Some trends elsewhere on the continent have shown cooling, while others show warming over the entire continent, but overall trends are smaller and dependent on season and the timespan over which the trend is computed. Climate models predict that temperature trends due to global warming will be much smaller in Antarctica than in the Arctic, mainly because heat uptake by the Southern Ocean acts to moderate the radiative forcing by greenhouse gases.

In a study released in 2009, historical weather station data was combined with satellite measurements to deduce past temperatures over large regions of the continent, and these temperatures indicate an overall warming trend. One of the paper's authors, Eric J. Steig of the University of Washington, stated "We now see warming is taking place on all seven of the earth's continents in accord with what models predict as a response to greenhouse gases." A follow-up study by O'Donnell and others that strongly criticized the Steig et al. work nevertheless found significant warming in West Antarctica. O'Donnell et al. also confirmed that Antarctica overall has been warming since the 1950s, but disagreed with Steig et al. about the strength of that warming. Subsequent measurements of temperatures in a borehole at the center of the West Antarctic ice sheet, by Orsi and others, found even larger positive trends than Steig et al.

Origin of the Controversy

Michael Crichton, in his 2004 novel *State of Fear*, asserted that cooling observed in the interior of the Antarctica shows the lack of reliability of the models used for global warming predictions, and thus of climate theory in general. This novel has a docudrama plot based upon the idea that there is a deliberately alarmist conspiracy behind global warming activism. As presented in page 193 of the novel: "The data show that one relatively small area called the Antarctic Peninsula is melting and calving huge icebergs. That's what gets reported year after year. But the continent as a whole is getting colder, and the ice is getting thicker." Other sources then picked up the argument, labeling it the "Antarctic Cooling Controversy", despite the fact that the small and variable observed trends are broadly consistent with the small magnitude of model-predicted temperature trends for Antarctica.

Crichton footnoted his assertion of Antarctic cooling as originating from the paper Doran *et al.*, 2002, although the paper referenced did not directly state that their measurements was evidence against global warming. The work stated: "Although previous reports suggest slight recent continental warming our spatial analysis of Antarctic meteorological data demonstrates a net cooling on the Antarctic continent between 1966 and 2000, particularly during summer and autumn. The McMurdo Dry Valleys have cooled by 0.7 °C per decade between 1986 and 2000, with similar pronounced seasonal trends.... Continental Antarctic cooling, especially the seasonality of cooling, poses challenges to models of climate and ecosystem change.

In response to Crichton, the lead author of the research paper, Peter Doran, published a statement in *The New York Times* stating, "... our results have been misused as 'evidence' against global warming by Michael Crichton in his novel *State of Fear*.... Our study did find that 58 percent of Antarctica cooled from 1966 to 2000. But during that period, the rest of the continent was warming. And climate models created since our paper was published have suggested a link between the lack of significant warming in Antarctica and the ozone hole over that continent. These models, conspicuously missing from the warming-skeptic literature, suggest that as the ozone hole heals — thanks to worldwide bans on ozone-destroying chemicals — all of Antarctica is likely to warm with the rest of the planet. An inconvenient truth?" He also emphasized the need for more stations in the Antarctic continent in order to obtain more robust results.

A rebuttal to Crichton's claims was presented by the group Real Climate:

> Long term temperature data from the Southern Hemisphere are hard to find, and by the time you get to the Antarctic continent, the data are extremely sparse. Nonetheless, some patterns do emerge from the limited data available. The Antarctic Peninsula, site of the now-defunct Larsen-B ice shelf, has warmed substantially. On the other hand, the few stations on the continent and in the interior appear to have cooled slightly (Doran et al., 2002; GISTEMP).
>
> At first glance this seems to contradict the idea of "global" warming, but one needs to be careful before jumping to this conclusion. A rise in the global mean temperature does not imply universal warming. Dynamical effects (changes in the winds and ocean circulation) can have just as large an impact, locally as the radiative forcing from greenhouse gases. The temperature change in any particular region will in fact be a combination of radiation-related changes (through greenhouse gases, aerosols, ozone and the like) and dynamical effects. Since the winds tend to only move heat from one place to another, their impact will tend to cancel out in the global mean.

It is common to find statements that "climate models generally predict amplified warming in polar regions" (*e.g.*, Doran *et al.*), a phenomenon called polar amplification. In fact, however, Arctic and Antarctic climates are out of phase with each other (the "polar see-saw" effect), and climate models predict amplified warming primarily for the Arctic and not for Antarctica.

Observations of Trends

There are few long term weather observations for Antarctica. There are less than twenty permanent stations in all and only two in the interior. More recently AWSs supplement this, but their records are relatively brief. Hence calculation of a trend for the entire continent is difficult. Satellite observations only exist since 1981 and provide surface temperature measurements only in cloud-free conditions.

The 2007 IPCC Fourth Assessment Report states, "Observational studies have presented evidence of pronounced warming over the Antarctic Peninsula, but little change over the rest of the continent during the last half of the 20th century." Chapman and Walsh note that "Trends calculated for the 1958–2002 period suggest modest warming over much of the 60°–90°S domain. All seasons show warming, with winter trends being the largest at +0.172 °C per decade while summer warming rates are only +0.045 °C per decade. The 45-year temperature trend for the annual means is +0.082 °C per decade corresponding to a +0.371 °C temperature change over the 1958–2002 period of record. Trends computed using these analyses show considerable sensitivity to start and end dates, with trends calculated using start dates prior to 1965 showing overall warming, while those using start dates from 1966 to 1982 show net cooling over the region."

Several scientific sources have reported that there is a cooling trend observed in the interior of the continent for the last two decades of the 20th century, while the Antarctic Peninsula shows a warming trend.

In early 2013, David Bromwich, a professor of polar meteorology at Ohio State University, and a team including Antarctic weather station experts from the University of Wisconsin, published a paper in *Nature Geoscience* showing that the warming in central West Antarctica was unambiguous—and likely about twice the magnitude estimated by Steig et al. The key to Bromwich et al.'s work was the correction for errors in the temperature sensors used in various incarnations of the Byrd Station record (the only long record in this part of Antarctica); miscalibration had previously caused the magnitude of the 1990s warmth to be underestimated, and the magnitude of the 2000s to be overestimated. The revised Byrd Station record is in very good agreement with the borehole temperature data from nearby WAIS Divide. A new statistical reconstruction shows significant warming over all of West Antarctic in the annual mean, driven by significant warming over most of the region in winter and spring. Summer and fall trends, are insignificant except over the Antarctic Peninsula where they are widespread only in fall. These finding are in good agreement with the 2009 study in Nature, though in general the new results show greater warming in West Antarctica and less warming over East Antarctica as a whole. Nicholas and Bromwich argue that while the warming in East Antarctica is not statistically significant, it would be greater in magnitude if not for the ozone hole. There is no evidence that any significant region of Antarctic has been cooling over the long term, except in fall. In a 2016 paper, Turner and others point out that if one considers just the last ~18 years, the trend on the Antarctic Peninsula has been cooling. This is likely connected with tropical variability, perhaps associated with the phase of the Interdecadal Pacific Oscillation.

Scientific Sources and Interpretations

According to a NASA press release:

> "Across most of the continent and the surrounding Southern Ocean, temperatures climbed... The temperature increases were greater and more widespread in West Antarctica than in East Antarctica, where some areas showed little change or even a cooling trend. This variability in temperature patterns across Antarctica complicates the work of scientists who are trying to understand the relative influence of natural cycles and human-caused climate change in Antarctica."

As a complement to NASA's findings, the British Antarctic Survey, which has undertaken the majority of Britain's scientific research in the area, has the following positions:

- Ice makes polar climate sensitive by introducing a strong positive feedback loop.

- Melting of continental Antarctic ice could contribute to global sea level rise.

- Climate models predict more snowfall than ice melting during the next 50 years, but models are not good enough for them to be confident about the prediction.

- Antarctica seems to be both warming around the edges and cooling at the center at the same time. Thus it is not possible to say whether it is warming or cooling overall.

- There is no evidence for a decline in overall Antarctic sea ice extent.

- The central and southern parts of the west coast of the Antarctic Peninsula have warmed by nearly 3 °C. The cause is not known.

- Changes have occurred in the upper atmosphere over Antarctica.

Research by Thompson and Solomon (2002) and by Shindell and Schmidt (2004) provide explanations for the observed cooling trend during the 1970s through 2000. An updated paper by Thompson et al. (2012) emphasized that this explanation only applies to austral summer; during the fall, winter and spring seasons, the mean trend is warming, and this is believed to be largely due to changes in atmospheric circulation related to warming trends in the tropical Pacific region.

Climate Change in the Arctic

The effects of global warming in the Arctic, or climate change in the Arctic include rising temperatures, loss of sea ice, and melting of the Greenland ice sheet with a related cold temperature anomaly, observed in recent years. Potential methane release from the region, especially through the thawing of permafrost and methane clathrates, is also a concern. The Arctic warms twice as fast compared to the rest of the world. The pronounced warming signal, the amplified response of the Arctic to global warming, it is often seen as a leading indicator of global warming. The melting of Greenland's ice sheet is linked to polar amplification. According to a study published in 2016, about 0.5°C of the warming in the Arctic has been attributed to reductions in sulfate aerosols in Europe since 1980.

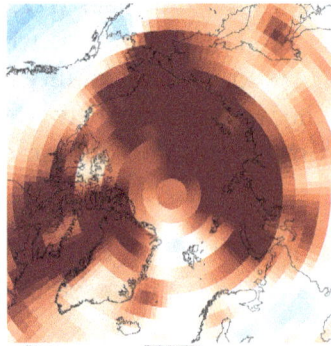

The image above shows where average air temperatures (October 2010 – September 2011) were up to 2 degrees Celsius above (red) or below (blue) the long-term average (1981–2010).

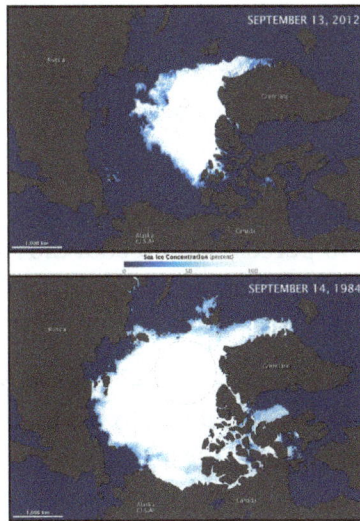

The maps above compare the Arctic ice minimum extents from 2012 (top) and 1984 (bottom). In 1984 the sea ice extent was roughly the average of the minimum from 1979 to 2000, and so was a typical year. The minimum sea ice extent in 2012 was roughly half of that average.

Rising Temperatures

According to the Intergovernmental Panel on Climate Change, "warming in the Arctic, as indicated by daily maximum and minimum temperatures, has been as great as in any other part of the world." The period of 1995–2005 was the warmest decade in the Arctic since at least the 17th century, with temperatures 2 °C (3.6 °F) above the 1951–1990 average. Some regions within the Arctic have warmed even more rapidly, with Alaska and western Canada's temperature rising by 3 to 4 °C (5.40 to 7.20 °F). This warming has been caused not only by the rise in greenhouse gas concentration, but also the deposition of soot on Arctic ice. A 2013 article published in Geophysical Research Letters has shown that temperatures in the region haven't been as high as they currently are since at least 44,000 years ago and perhaps as long as 120,000 years ago. The authors conclude that "anthropogenic increases in greenhouse gases have led to unprecedented regional warmth."

Arctic Amplification

The poles of the Earth are more sensitive to any change in the planet's climate than the rest of the planet. In the face of ongoing global warming, the poles are warming faster than lower latitudes. The primary cause of this phenomenon is ice-albedo feedback, whereby melting ice uncovers darker land or ocean beneath, which then absorbs more sunlight, causing more heating. The loss of the Arctic sea ice may represent a tipping point in global warming, when 'runaway' climate change starts, but on this point the science is not yet settled. According to a 2015 study, based on computer modelling of aerosols in the atmosphere, up to 0.5 degrees Celsius of the warming observed in the Arctic between 1980 and 2005 is due to aerosol reductions in Europe.

Black Carbon

Black carbon deposits (from the exhaust system of marine engines) reduce the albedo when deposited on snow and ice, and thus accelerate the effect of the melting of snow and sea ice.

According to a 2015 study, reductions in black carbon emissions and other minor greenhouse gases, by roughly 60 percent, could cool the Arctic up to 0.2 °C by 2050.

Decline of Sea Ice

Sea ice is currently in decline in area, extent, and volume and may cease to exist sometime during the 21st century. Sea ice area refers to the total area covered by ice, whereas sea ice extent is the area of ocean with at least 15% sea ice, while the volume is the total amount of ice in the Arctic.

Changes in Extent and Area

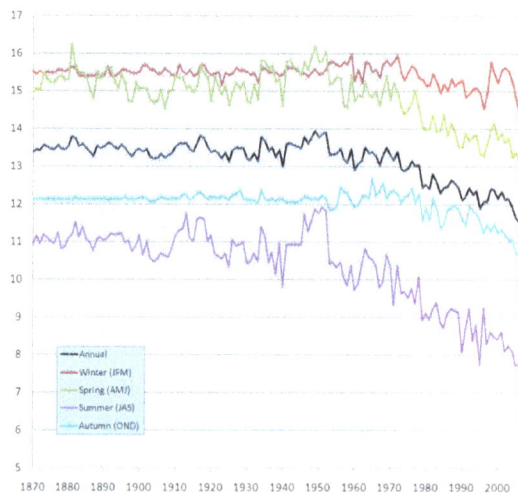

1870–2009 Northern Hemisphere sea ice extent in million square kilometers. Blue shading indicates the pre-satellite era; data then is less reliable. In particular, the near-constant level extent in autumn up to 1940 reflects lack of data rather than a real lack of variation.

Reliable measurement of sea ice edges began with the satellite era in the late 1970s. Before this time, sea ice area and extent were monitored less precisely by a combination of ships, buoys and aircraft. The data show a long-term negative trend in recent years, attributed to global warming, although there is also a considerable amount of variation from year to year. Some of this variation may be related to effects such as the arctic oscillation, which may itself be related to global warming.

The Arctic sea ice September minimum extent (i.e., area with at least 15% sea ice coverage) reached new record lows in 2002, 2005, 2007, and 2012. The 2007 melt season let to a minimum 39% below the 1979–2000 average, and for the first time in human memory, the fabled Northwest Passage opened completely. The dramatic 2007 melting surprised and concerned scientists.

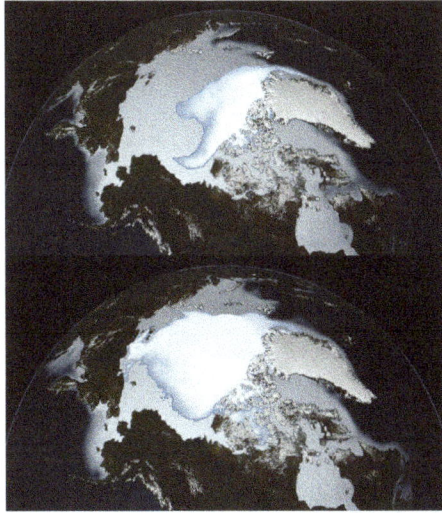

Sea ice coverage in 1980 (bottom) and 2012 (top), as observed by passive microwave sensors on NASA's Nimbus-7 satellite and by the Special Sensor Microwave Imager/Sounder (SSMIS) from the Defense Meteorological Satellite Program (DMSP). Multi-year ice is shown in bright white, while average sea ice cover is shown in light blue to milky white. The data shows the ice cover for the period of 1 November through 31 January in their respective years.

From 2008 to 2011, Arctic sea ice minimum extent was higher than 2007, but it did not return to the levels of previous years. In 2012 however, the 2007 record low was broken in late August with 3 weeks still left in the melt season. It continued to fall, bottoming out on 16 September 2012 at 3.41 million square kilometers (1.32 million square miles), or 760,000 square kilometers (293,000 square miles) below the previous low set on 18 September 2007 and 50% below the 1979–2000 average.

The rate of the decline in entire arctic ice coverage is accelerating. From 1979–1996, the average per decade decline in entire ice coverage was a 2.2% decline in ice extent and a 3% decline in ice area. For the decade ending 2008, these values have risen to 10.1% and 10.7%, respectively. These are comparable to the September to September loss rates in year-round ice (i.e., perennial ice, which survives throughout the year), which averaged a retreat of 10.2% and 11.4% per decade, respectively, for the period 1979–2007.

Changes in Volume

The sea ice thickness field and accordingly the ice volume and mass, is much more difficult to determine than the extension. Exact measurements can be made only at a limited number of points. Because of large variations in ice and snow thickness and consistency air- and spaceborne-measurements have to be evaluated carefully. Nevertheless, the studies made support the assumption of a dramatic decline in ice age and thickness. While the arctic ice area and extent show an accelerating downward trend, arctic ice volume shows an even sharper decline than the ice coverage. Since 1979, the ice volume has shrunk by 80% and in just the past decade the volume declined by 36% in the autumn and 9% in the winter.

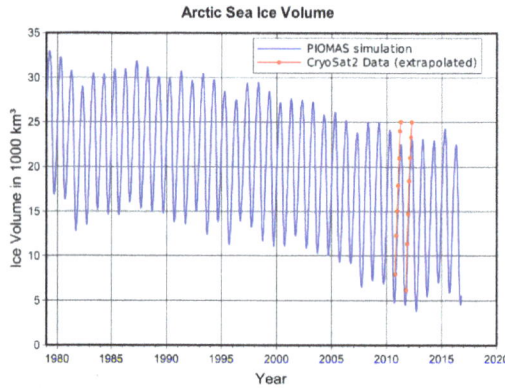

Seasonal variation and long term decrease of Arctic sea ice volume as determined by measurement backed numerical modelling.

An End to Summer Sea Ice

The IPCC's Fourth Assessment Report in 2007 summarized the current state of sea ice projections: "the projected reduction [in global sea ice cover] is accelerated in the Arctic, where some models project summer sea ice cover to disappear entirely in the high-emission A2 scenario in the latter part of the 21st century." However, current climate models frequently underestimate the rate of sea ice retreat. A summertime ice-free arctic would be unprecedented in recent geologic history, as currently scientific evidence does not indicate an ice-free polar sea anytime in the last 700,000 years.

The Arctic ocean will likely be free of summer sea ice before the year 2100, but many different dates have been projected. One study suggests 2060–2080, another 2030, and, yet another, 2016. A 2013 study showed that simply extending summertime ice melting trends into the future in a straight line predicts an ice-free summertime Arctic as early as by 2020.

Permafrost Thaw

This century, thawing of the various types of Arctic permafrost could release large amounts of carbon into the atmosphere. It has been estimated that about two-thirds of released carbon escapes to the atmosphere as carbon dioxide, originating primarily from ancient ice deposits along the ~7,000 kilometer long coastline of the East Siberian Arctic Shelf (ESAS) and shallow subsea permafrost. Following thaw, collapse and erosion of coastline and seafloor deposits may accelerate with Arctic amplification of climate warming.

Rapidly thawing Arctic permafrost and coastal erosion on the Beaufort Sea,
Arctic Ocean, near Point Lonely, AK.

Permafrost thaw ponds in Hudson Bay Canada near Greenland

Climate models suggest that during periods of rapid sea-ice loss, temperatures could increase as far as 1,450 km (900 mi) inland, accelerating the rate of terrestrial permafrost thaw, with consequential effects on carbon and methane release.

Subsea Permafrost

Subsea permafrost occurs beneath the seabed and exists in the continental shelves of the polar regions. This source of methane is different from methane clathrates, but contributes to the overall outcome and feedbacks.

Sea ice serves to stabilise methane deposits on and near the shoreline, preventing the clathrate breaking down and venting into the water column and eventually reaching the atmosphere. From sonar measurements in recent years researchers quantified the density of bubbles emanating from the subsea permafrost into the Ocean (a process called ebullition), and found that 100–630 mg methane per square meters is emitted daily along the East Siberian Shelf, into the water column. They also found that during storms, methane levels in the water column drop dramatically, when wind driven air-sea gas exchange accelerates the ebullition process into the atmosphere. This observed pathway suggest that methane from seabed permafrost will progress rather slowly, instead of abrupt changes. However, Arctic cyclones, fueled by global warming and further accumulation of greenhouse gases in the atmosphere could contribute to more release from this methane cache, which is really important for the Artic.

Changes in Vegetation

Bloody Falls in July 2007

Changes in vegetation are associated with the increases in landscape scale methane emissions.

The growing season has lengthened in the far northern latitudes, bringing major changes to plant communities in tundra and boreal (also known as taiga) ecosystems.

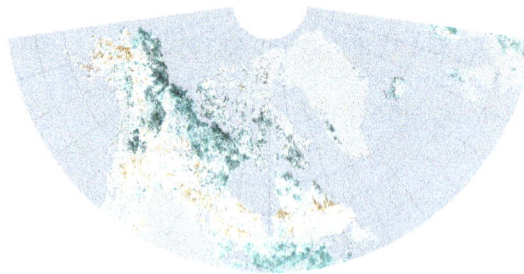

Western Hemisphere Arctic Vegetation Index Trend

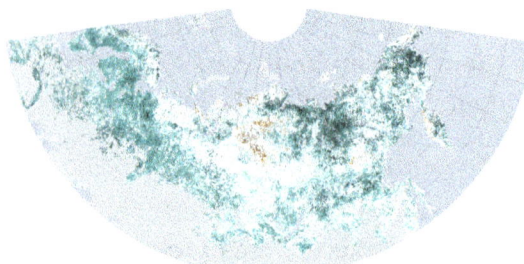

Eastern Hemisphere Vegetation Index Trend

For decades, NASA and NOAA satellites have continuously monitored vegetation from space. The Moderate Resolution Imaging Spectroradiometer (MODIS) and Advanced Very High Resolution Radiometer (AVHRR) instruments measure the intensity of visible and near-infrared light reflecting off of plant leaves. Scientists use the information to calculate the Normalized Difference Vegetation Index (NDVI), an indicator of photosynthetic activity or "greenness" of the landscape.

The maps above show the Arctic Vegetation Index Trend between July 1982 and December 2011 in the Arctic Circle. Shades of green depict areas where plant productivity and abundance increased; shades of brown show where photosynthetic activity declined. The maps show a ring of greening in the treeless tundra ecosystems of the circumpolar Arctic—the northernmost parts of Canada, Russia, and Scandinavia. Tall shrubs and trees started to grow in areas that were previously dominated by tundra grasses. The researchers concluded that plant growth had increased by 7 to 10 percent overall.

However, boreal forests, particularly those in North America, showed a different response to warming. Many boreal forests greened, but the trend was not as strong as it was for tundra of the circumpolar Arctic. In North America, some boreal forests actually experienced "browning" (less photosynthetic activity) over the study period. Droughts, forest fire activity, animal and insect behavior, industrial pollution, and a number of other factors may have contributed to the browning.

"Satellite data identify areas in the boreal zone that are warmer and drier and other areas that are warmer and wetter," explained co-author Ramakrishna Nemani of NASA's Ames Research Center. "Only the warmer and wetter areas support more growth."

"We found more plant growth in the boreal zone from 1982 to 1992 than from 1992 to 2011, because water limitations were encountered in the later two decades of our study," added co-author Sangram Ganguly of the Bay Area Environmental Research Institute and NASA Ames.

The less severe winters in tundra areas allow shrubs such as alders and dwarf birch to replace moss and lichens. The impact on mosses and lichens is unclear as there exist very few studies at species level, also climate change is more likely to cause increased fluctuation and more frequent extreme events. The feedback effect of shrubs on the tundra's permafrost is unclear. In the winter they trap more snow which insulates the permafrost from extreme cold spells, but in the summer they shade the ground from direct sunlight. The warming is likely to cause changes in the plant communities. Except for an increase in shurbs, warming may also cause a decline in cushion plants such as moss campion. Since cushion plants act as facilitator species across trophic level and fill important roles in severe environments this could cause cascading effects in the ecosystems. Rising summer temperature melts on Canada's Baffin Island have revealed moss previously covered which has not seen daylight in 44,000 years.

The reduction of sea ice has boosted the productivity of phytoplankton by about twenty percent over the past thirty years. However, the effect on marine ecosystems is unclear, since the larger types of phytoplankton, which are the preferred food source of most zooplankton, do not appear to have increased as much as the smaller types. So far, arctic phytoplankton have not had a significant impact on the global carbon cycle. In summer, the melt ponds on young and thin ice have allowed sunlight to penetrate the ice, in turn allowing phytoplankton to bloom in unexpected concentrations, although it is unknown just how long this phenomenon has been occurring.

Changes for Animals

The northward shift of the subarctic climate zone is allowing animals that are adapted to that climate to move into the far north, where they are replacing species that are more adapted to a pure arctic climate. Where the arctic species are not being replaced outright, they are often interbreeding with their southern relations. Among slow-breeding vertebrate species, this often has the effect of reducing the genetic diversity of the genus. Another concern is the spread of infectious diseases, such as brucellosis or phocine distemper virus, to previously untouched populations. This is a particular danger among marine mammals who were previously segregated by sea ice.

3 April 2007, the National Wildlife Federation urged the United States Congress to place polar bears under the Endangered Species Act. Four months later, the United States Geological Survey completed a year-long study which concluded in part that the floating Arctic sea ice will continue its rapid shrinkage over the next 50 years, consequently wiping out much of the polar bear habitat. The bears would disappear from Alaska, but would continue to exist in the Canadian Arctic Archipelago and areas off the northern Greenland coast. Secondary ecological effects are also resultant from the shrinkage of sea ice; for example, polar bears are denied their historic length of seal hunting season due to late formation and early thaw of pack ice.

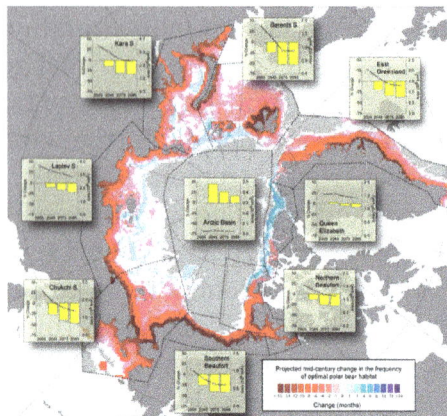

Projected change in polar bear habitat from 2001–2010 to 2041–2050.

Melting of the Greenland Ice Sheet

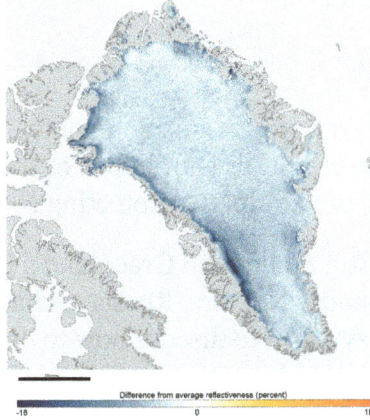

Albedo Change on Greenland

Models predict a sea-level contribution of about 5 centimetres (2 in) from melting in Greenland during the 21st century. It is also predicted that Greenland will become warm enough by 2100 to begin an almost complete melt during the next 1,000 years or more. In early July 2012, 97% percent of the Ice Sheet experienced some form of surface melt including the summits.

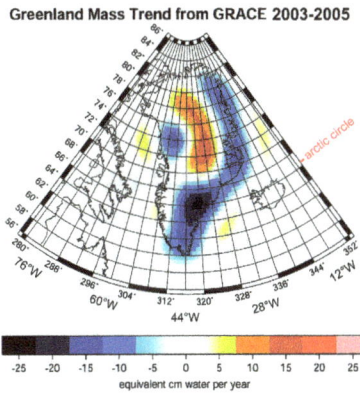

Greenland Ice Sheet Mass Trend 2003–2005.

Ice thickness measurements from the GRACE satellite indicate that ice mass loss is accelerating. For the period 2002–2009, the rate of loss increased from −137 Gt/yr to −286 Gt/yr, with an acceleration of −30 gigatonnes per year per year.

Effect on Ocean Circulation

Although this is now thought unlikely in the near future, it has also been suggested that there could be a shutdown of thermohaline circulation, similar to that which is believed to have driven the Younger Dryas, an abrupt climate change event. There is also potentially a possibility of a more general disruption of ocean circulation, which may lead to an ocean anoxic event, although these are believed to be much more common in the distant past. It is unclear whether the appropriate pre-conditions for such an event exist today.

Territorial Claims

Growing evidence that global warming is shrinking polar ice has added to the urgency of several nations' Arctic territorial claims in hopes of establishing resource development and new shipping lanes, in addition to protecting sovereign rights.

Danish Foreign Minister Per Stig Møller and Greenland's Premier Hans Enoksen invited foreign ministers from Canada, Norway, Russia and the United States to Ilulissat, Greenland for a summit in May 2008 to discuss how to divide borders in the changing Arctic region, and a discussion on more cooperation against climate change affecting the Arctic. At the Arctic Ocean Conference, Foreign Ministers and other officials representing the five countries announced the Ilulissat Declaration on 28 May 2008.

Social Impacts

People are affecting the geographic space of the Arctic and the Arctic is affecting the population. Much of the climate change in the Arctic can be attributed to humans influences on the atmosphere, such as an increased greenhouse effect caused by the increase in CO_2 due to the burning of fossil fuels. Climate change is having a direct impact on the people that live in the Arctic, as well as other societies around the world.

The warming environment presents challenges to local communities such as the Inuit. Hunting, which is a major way of survival for some small communities, will be changed with increasing temperatures. The reduction of sea ice will cause certain species populations to decline or even become extinct. In good years, some communities are fully employed by the commercial harvest of certain animals. The harvest of different animals fluctuates each year and with the rise of temperatures it is likely to continue changing and creating issues for Inuit hunters. Unsuspected changes in river and snow conditions will cause herds of animals, including reindeer, to change migration patterns, calving grounds, and forage availability.

Other forms of transportation in the Arctic have seen negative impacts from the current warming, with some transportation routes and pipelines on land being disrupted by the melting of ice. Many Arctic communities rely on frozen roadways to transport supplies and travel from area to area. The changing landscape and unpredictability of weather is creating new challenges in the Arctic.

Navigation

The Transpolar Sea Route is a future Arctic shipping lane running from the Atlantic Ocean to the Pacific Ocean across the center of the Arctic Ocean. The route is also sometimes called Trans-Arctic Route. In contrast to the Northeast Passage (including the Northern Sea Route) and the North-West Passage it largely avoids the territorial waters of Arctic states and lies in international high seas.

Governments and private industry have shown a growing interest in the Arctic. Major new shipping lanes are opening up: the northern sea route had 34 passages in 2011 while the Northwest Passage had 22 traverses, more than any time in history. Shipping companies may benefit from the shortened distance of these northern routes. Access to natural resources will increase, including valuable minerals and offshore oil and gas. Finding and controlling these resources will be difficult with the continually moving ice. Tourism may also increase as less sea ice will improve safety and accessibility to the Arctic.

The melting of Arctic ice caps is likely to increase traffic in and the commercial viability of the Northern Sea Route. One study, for instance, projects, "remarkable shifts in trade flows between Asia and Europe, diversion of trade within Europe, heavy shipping traffic in the Arctic and a substantial drop in Suez traffic. Projected shifts in trade also imply substantial pressure on an already threatened Arctic ecosystem".

Research

National

Individual countries within the Arctic zone, Canada, Denmark (Greenland), Finland, Iceland, Norway, Russia, Sweden, and the United States (Alaska) conduct independent research through a variety of organizations and agencies, public and private, such as Russia's Arctic and Antarctic Research Institute. Countries who do not have Arctic claims, but are close neighbors, conduct Arctic research as well, such as the Chinese Arctic and Antarctic Administration (CAA). The United States's National Oceanic and Atmospheric Administration (NOAA) produces an Arctic Report Card annually, containing peer-reviewed information on recent observations of environmental conditions in the Arctic relative to historical records.

International

International cooperative research between nations has become increasingly important:

- Arctic climate change is summarized by the Intergovernmental Panel on Climate Change (IPCC) in its series of Assessment Reports and the Arctic Climate Impact Assessment.

- DAMOCLES (Developing Arctic Modeling and Observing Capabilities for Long-term Environmental Studies): European integrated project "specifically concerned with the potential for a significantly reduced sea ice cover, and the impacts this might have on the environment and on human activities, both regionally and globally".

- European Space Agency (ESA) launched CryoSat-2 on 8 April 2010. It provides satellite data on Arctic ice cover change rates.

- International Arctic Buoy Program: deploys and maintains buoys that provide real-time position, pressure, temperature, and interpolated ice velocity data.

- International Arctic Research Center: Main participants are the United States and Japan.

- International Arctic Science Committee: non-governmental organization (NGO) with diverse membership, including 18 countries from 3 continents.

- 'Role of the Arctic Region', in conjunction with the International Polar Year, was the focus of the second international conference on Global Change Research, held in Nynäshamn, Sweden, October 2007.

- SEARCH (Study of Environmental Arctic Change): Supported by the Arctic Research Office, a division of the United States' National Oceanic and Atmospheric Administration (NOAA), and the Russian Academy of Sciences.

References

- Goldenberg S (24 July 2012). "Greenland ice sheet melted at unprecedented rate during July". The Guardian. London. Retrieved 4 November 2012

- Elizabeth Peacock; Mitchell K. Taylor; Jeffrey Laake; Ian Stirling (April 2013). "Population ecology of polar bears in Davis Strait, Canada and Greenland". The Journal of Wildlife Management. 77 (3): 463–476. doi:10.1002/jwmg.489. Retrieved January 26, 2014

- National Research Council (2010). America's Climate Choices: Panel on Advancing the Science of Climate Change;. Washington, D.C.: The National Academies Press. ISBN 0-309-14588-0. Archived from the original on 29 May 2014

- USGCRP (2015), Glossary, Washington, DC, USA: U.S. Global Change Research Program (USGCRP), retrieved 20 January 2014

- J. C. Stroeve; T. Markus; L. Boisvert; J. Miller; A. Barrett (2014). "Changes in Arctic melt season and implications for sea ice loss". Bibcode:2014GeoRL..41.1216S. doi:10.1002/2013GL058951

- William L. Chapman; John E. Walsh (2007). "A Synthesis of Antarctic Temperatures". Journal of Climate. 20 (16): 4096–4117. Bibcode:2007JCli...20.4096C. doi:10.1175/JCLI4236.1. Retrieved 2007-11-05

- IPCC AR4 SYR (2007), Core Writing Team; Pachauri, R.K; Reisinger, A., eds., Climate Change 2007: Synthesis Report, Contribution of Working Groups I, II and III to the Fourth Assessment Report of the Intergovernmental Panel on Climate Change, Geneva, Switzerland: IPCC, ISBN 92-9169-122-4

- IPCC (November 2010), Understanding Climate Change: 22 years of IPCC assessment (PDF), IPCC, Archived from the original on 22 September 2014, retrieved 29 May 2014

- Jahn, A.; Kay, J.E.; Holland, M.M.; Hall, D.M. (2016). "How predictable is the timing of a summer ice-free Arctic?". Geophysical Research Letters. 43. doi:10.1002/2016GL070067

- Josefino C. Comiso (2000). "Variability and Trends in Antarctic Surface Temperatures from In Situ and Satellite Infrared Measurements" (PDF). Journal of Climate. 13 (10): 1674–1696. doi:10.1175/1520-0442(2000)013. Retrieved 2008-08-14

- Karl, Thomas R.; Melillo, Jerry M.; Peterson, Thomas C., eds. (2009). Global Climate Change Impacts in the United States (PDF). New York: Cambridge University Press. ISBN 978-0-521-14407-0

- UKMO (18 September 2013), AVOID Reports, UK Meteorological Office (UKMO), Archived from the original on 8 June 2014, retrieved 8 June 2014

- "Arctic cut-off high drives the poleward shift of a new Greenland melting record". Nature Communications. 7: 11723. doi:10.1038/ncomms11723

- Bekkers, Eddy; Francois, Joseph F.; Rojas-Romagosa, Hugo (2016-12-01). "Melting Ice Caps and the Economic Impact of Opening the Northern Sea Route". The Economic Journal: n/a–n/a. ISSN 1468-0297. doi:10.1111/ecoj.12460

- US NRC (2010). Advancing the Science of Climate Change. A report by the US National Research Council (NRC). Washington, D.C., USA: National Academies Press. ISBN 0-309-14588-0. Archived from the original on 29 May 2014

- Maslowski, Wieslaw (16 March 2010). "Advancements and Limitations in Understanding and Predicting Arctic Climate Change". State of the Arctic (conference website). Retrieved 2 February 2015

- Velicogna, I. (2009). "Increasing rates of ice mass loss from the Greenland and Antarctic ice sheets revealed by GRACE". Geophysical Research Letters. 36: L19503. Bibcode:2009GeoRL..3619503V. doi:10.1029/2009GL040222

- Peter N. Spotts (2002-01-18). "Guess what? Antarctica's getting colder, not warmer". The Christian Science Monitor. Retrieved 2013-04-13

PERMISSIONS

All chapters in this book are published with permission under the Creative Commons Attribution Share Alike License or equivalent. Every chapter published in this book has been scrutinized by our experts. Their significance has been extensively debated. The topics covered herein carry significant information for a comprehensive understanding. They may even be implemented as practical applications or may be referred to as a beginning point for further studies.

We would like to thank the editorial team for lending their expertise to make the book truly unique. They have played a crucial role in the development of this book. Without their invaluable contributions this book wouldn't have been possible. They have made vital efforts to compile up to date information on the varied aspects of this subject to make this book a valuable addition to the collection of many professionals and students.

This book was conceptualized with the vision of imparting up-to-date and integrated information in this field. To ensure the same, a matchless editorial board was set up. Every individual on the board went through rigorous rounds of assessment to prove their worth. After which they invested a large part of their time researching and compiling the most relevant data for our readers.

The editorial board has been involved in producing this book since its inception. They have spent rigorous hours researching and exploring the diverse topics which have resulted in the successful publishing of this book. They have passed on their knowledge of decades through this book. To expedite this challenging task, the publisher supported the team at every step. A small team of assistant editors was also appointed to further simplify the editing procedure and attain best results for the readers.

Apart from the editorial board, the designing team has also invested a significant amount of their time in understanding the subject and creating the most relevant covers. They scrutinized every image to scout for the most suitable representation of the subject and create an appropriate cover for the book.

The publishing team has been an ardent support to the editorial, designing and production team. Their endless efforts to recruit the best for this project, has resulted in the accomplishment of this book. They are a veteran in the field of academics and their pool of knowledge is as vast as their experience in printing. Their expertise and guidance has proved useful at every step. Their uncompromising quality standards have made this book an exceptional effort. Their encouragement from time to time has been an inspiration for everyone.

The publisher and the editorial board hope that this book will prove to be a valuable piece of knowledge for students, practitioners and scholars across the globe.

Index